Host Immune Responses and Pathogenesis to *Brucella* spp. Infection

Host Immune Responses and Pathogenesis to *Brucella* spp. Infection

Editors

Sergio Costa Oliveira
Guillermo Giambartolomei

MDPI • Basel • Beijing • Wuhan • Barcelona • Belgrade • Manchester • Tokyo • Cluj • Tianjin

Editors

Sergio Costa Oliveira
Universidade Federal de Minas Gerais
Brazil

Guillermo Giambartolomei
Universidad de Buenos Aires
Argentina

Editorial Office
MDPI
St. Alban-Anlage 66
4052 Basel, Switzerland

This is a reprint of articles from the Special Issue published online in the open access journal *Pathogens* (ISSN 2076-0817) (available at: https://www.mdpi.com/journal/pathogens/special_issues/host_immune_responses_and_pathogenesis_to_brucella_spp_infection).

For citation purposes, cite each article independently as indicated on the article page online and as indicated below:

LastName, A.A.; LastName, B.B.; LastName, C.C. Article Title. *Journal Name* **Year**, *Volume Number*, Page Range.

ISBN 978-3-0365-2418-4 (Hbk)
ISBN 978-3-0365-2419-1 (PDF)

Cover image courtesy of Sergio Costa Oliveira

Contents

About the Editors

Sergio Costa Oliveira is Professor in Immunology at the Department of Biochemistry and Immunology of the Federal University of Minas Gerais (UFMG), where he is the Head of the Laboratory of Immunology of Infectious Diseases composed of 20 members from among graduate students, post-docs, and technicians. In 2009, Prof. Oliveira was elected member of the Brazilian Academy of Science. He has over 220 peer-reviewed publications with more than 6000 citations. Prof. Sergio C. Oliveira is internationally recognized for his scientific accomplishments studying the host immune response against the bacterial pathogen Brucella abortus, agent of the worldwide zoonosis termed brucellosis.

Guillermo Giambartolomei, Biochemist, earned his Ph.D. degree in Immunology at the University of Buenos Aires and performed his postdoctoral training at Tulane University, USA. He is Assistant Professor of Immunology at the School of Medicine at the University of Buenos Aires (UBA) and Principal Investigator at the National Council of Research of Argentina (CONICET). He is the Head of the laboratory of Immunity and Immunopathology against intracellular bacteria at the Institute of Immunology, Genetics and Metabolism (INIGEM) (CONICET-UBA) and acts as Vice-Director of INIGEM. In 2014, he acted as President of the Argentine Society of Immunology (SAI). His areas of interest in the field of brucellosis are innate immunity, bacterial determinants of immunopathology, and mechanisms of immune escape.

pathogens

MDPI

Editorial

Host Immune Responses and Pathogenesis to *Brucella* spp. Infection

Sergio C. Oliveira

Department of Biochemistry and Immunology, Institute of Biological Sciences, Federal University of Minas Gerais, Belo Horizonte, Minas Gerais 31270-901, Brazil; scozeus1@gmail.com

check for updates

Citation: Oliveira, S.C. Host Immune Responses and Pathogenesis to *Brucella* spp. Infection. *Pathogens* **2021**, *10*, 288. https://doi.org/10.3390/pathogens10030288

Received: 20 February 2021
Accepted: 22 February 2021
Published: 3 March 2021

Brucellosis, caused by the facultative intracellular bacteria *Brucella* species, is one the most prevalent zoonoses worldwide. *Brucella* causes >500,000 human infections per year, and brucellosis is underreported in endemic areas [1]. Between livestock losses and human morbidity, brucellosis imposes significant economic impacts, perpetuating poverty in endemic regions [2]. There is a considerable amount of evidence that indicates the capacity of *Brucella* sp. to avoid or interfere with components of the host immune responses, which plays a critical role in their virulence. It has been suggested that *Brucella* has developed a stealth strategy through pathogen-associated molecular patterns reduction, modification, and hiding to ensure low stimulatory activity and toxicity for cells [3]. This strategy allows *Brucella* to reach its replication niche before activating antimicrobial mechanisms by host immune responses. However, inside the host cells, *Brucella* releases vital molecules for the bacteria that trigger the activation of host cytosolic receptors [4,5]. In the paper by Tupik et al. [6], in vivo studies using $Asc^{-/-}$ mice infected with *Brucella* revealed an increased bacteria load and decreased immune cell recruitment. The findings of the study suggest that the protective role of ASC may result from the induction of pyroptosis through a gasdermin D-dependent mechanism in macrophages. However, further studies are required to elucidate this complex circuit by which the host immune system recognizes *Brucella*-derived molecules. This editorial summarizes the data described in the Special Issue entitled "Host Immune Response and Pathogenesis to *Brucella* spp. Infection" consisting of seven research articles and two reviews. These contributions report several aspects of host–*Brucella* interactions, and these findings will help to advance the comprehension of bacterial pathogenesis and contribute to the future development of drugs or vaccines to control brucellosis.

For a successful infection process, a pathogen has to invade, survive, and replicate in host cells. During the first steps of *Brucella* trafficking, the bacteria is able to block the progression of its cell cycle, remaining at the G1 stage for several hours, before it reaches its replication niche. The work of Van der Henst et al. [7] demonstrated that starvation mediated by guanosine tetra- or penta-phosphate, (p)ppGpp, is one of the factors contributing to G1 arrest observed in *B. abortus* infection in macrophages. Adhesion to target cells is another major step forward for bacterial invasion and replication. Bialer et al. [8] reviewed the *Brucella* adhesins and their role in mediating adhesion to cells. These molecules include the sialic acid-binding proteins SP29 and SP41 (binding to erythrocytes and epithelial cells, respectively), the BigA and BigB proteins that contain an Ig-like domain (binding to cell adhesion molecules in epithelial cells), the monomeric autotransporters BmaA, BmaB, and BmaC (binding to extracellular matrix components, epithelial cells, osteoblasts, synoviocytes, and trophoblasts), the trimeric autotransporters BtaE and BtaF (binding to ECM components and epithelial cells), and Bp26 (binding to ECM components). After binding, replication in phagocytic and non-phagocytic cells is required to establish infection. One of the main clinical signs of brucellosis is abortion in domestic animals. Zavattieri et al. [9] demonstrate that *Brucella abortus* was able to infect and survive in both non-decidualized and decidualized human endometrial stromal cells (T-HESC cell line). *Brucella* infection did not induce cytotoxicity and did not alter the decidualization status of cells, but elicited the secretion of IL-8 and MCP-1 in either decidualized or non-decidualized T-HESC. The proinflammatory responses induced by *Brucella* infection in T-HESC may contribute to the gestational complications and abortion during brucellosis.

Brucellae reside mostly within phagocytes and other cells, including trophoblasts, where they establish a preferred replicative niche inside the endoplasmic reticulum. González-Espinoza et al. [10] propose that *Brucella* takes advantage of the environment provided by the cellular niches in which it resides to generate reservoirs and disseminate to other organs, such as spleen, lymph nodes, liver, bone marrow, epididymis, and placenta. They discuss in this review how the favored cellular niches for *Brucella* infection in the host give rise to anatomical reservoirs that may lead to chronic infections or persistence in asymptomatic subjects, and which may be considered a threat for further contamination. The natural infection by *Brucella* occurs mainly by oral and nasal routes through the consumption of raw milk and unpasteurized dairy products from infected animals, the inhalation of aerosols containing the pathogen, and/or contact with infected animals and their secretions. Considering that the oral route is the main route of natural infection in humans and animals, there is a need to understand the mechanisms of the establishment of oral infection so that new therapeutic strategies can be developed in order to control this disease. Santos et al. [11] report the role of ST2 receptor in a murine model of oral infection with *Brucella abortus* and its influence on gut homeostasis and control of bacterial replication. Their results suggest that $ST2^{-/-}$ are more resistant to *B. abortus* infection, as lower bacterial CFUs were detected in the livers and spleens of knockout mice when compared to wild-type Additionally, they observed an increase in intestinal permeability in WT infected mice compared to $ST2^{-/-}$ animals. Finally, their findings suggest that ST2 receptor is involved in the invasion process of *B. abortus* by the mucosa in the oral infection model.

Regarding pathology, the most frequent clinical characteristics of brucellosis besides abortion are hepatomegaly, splenomegaly, and peripheral lymphadenopathy, revealing the preference of *Brucella* for the reticuloendothelial system. The research article by Arriola-Benitez et al. [12] describes how *Brucella abortus* infection induces the upregulation of class II transactivator protein (CIITA) with concomitant MHC-I and -II expression in immortalized human hepatic stellate cell line (LX-2) in a manner that is independent from the expression of the type 4 secretion system (T4SS). Since hepatocytes constitute the most abundant epithelial cell in the liver, experiments were conducted to determine the contribution of these cells in antigen presentation in the context of *B. abortus* infection. The results indicated that *B. abortus*-infected hepatocytes have an increased MHC-I expression, but MHC-II levels remain at basal levels. Overall, the authors revealed that *B. abortus* infection of hepatic stellate cells and hepatocytes is able to differentially regulate the MHC expression, thus stimulating the T-cell specific-immune response in the liver. The central nervous system (CNS) invasion by bacteria of the genus *Brucella* results in an inflammatory disorder termed neurobrucellosis. The precise mechanism whereby the bacterium leaves the bloodstream and gains access to the CNS remains unclear. Regardless of the mechanism, it is clear that once the bacterium reaches the CNS it induces a pathological pro-inflammatory response. The study of Rodriguez et al. [13] investigated the role of *Brucella abortus*-stimulated platelets on human brain microvascular endothelial cell (HBMEC) activation. Platelets enhanced HBMEC activation in response to *B. abortus* infection. Additionally, supernatants from *B. abortus*-activated platelets promoted the transendothelial migration of neutrophils and monocytes depending on the Erk1/2 signaling pathway. The results of this study describe a mechanism whereby *B. abortus*-stimulated platelets induce endothelial cell activation, promoting neutrophils and monocytes to traverse the blood–brain barrier, probably contributing to the inflammatory pathology of neurobrucellosis.

Vaccination is the major countermeasure to control *Brucella* infection. Currently used *Brucella* vaccines, *Brucella abortus* strain 19 and RB51, are comprised of live attenuated

Brucella strains and prevent infection in animals. The study of Gupta et al. [14] tested the recombinant proteins Omp25 and L7/L12 as potential vaccine candidates. Challenge with virulent *B. abortus* 544 demonstrated that Omp25+L7/L12-vaccinated mice exhibited superior log10 protection (1.98) compared to individual vaccines L7/L12 (1.75) and Omp25 (1.46). However, further studies are necessary to test the effectiveness of this divalent vaccine in large animals. Additionally, the cost of recombinant vaccines has to be taken into account when discussing veterinary vaccines. The articles included in this Special Issue present novel data on *Brucella* spp. infections, including host immune responses and bacterial pathogenesis that contribute significantly to improving the understanding of this disease.

Funding: This study was carried out with the financial support of CNPq (grants# 302660/2015-1 and 303044/2020-9) and the National Institute of Health R01 AI116453.

Acknowledgments: I would like to thank all of the authors of the papers published in this Special Issue.

Conflicts of Interest: The author declares no conflict of interest.

References

1. Pappas, G.; Papadimitriou, P.; Akritidis, N.; Christou, L.; Tsianos, E.V. The new global map of human brucellosis. *Lancet Infect. Dis.* **2006**, *6*, 91–99. [CrossRef]
2. Rossetti, C.A.; Arenas-Gamboa, A.M.; Maurizio, E. Caprine brucellosis: A historically neglected disease with significant impact on public health. *PLoS Negl. Trop. Dis.* **2017**, *11*, e0005692. [CrossRef] [PubMed]
3. Marim, F.M.; Franco, M.M.C.; Gomes, M.T.R.; Miraglia, M.C.; Giambartolomei, G.H.; Oliveira, S.C. The role of NLRP3 and AIM2 in inflammasome activation during Brucella abortus infection. *Semin. Immunopathol.* **2017**, *39*, 215–223. [CrossRef] [PubMed]
4. Gomes, M.T.; Campos, P.C.; Oliveira, F.S.; Corsetti, P.P.; Bortoluci, K.R.; Cunha, L.D.; Zamboni, D.S.; Oliveira, S.C. Critical role of ASC inflammasomes and bacterial type IV secretion system in caspase-1 activation and host innate resistance to Brucella abortus infection. *J. Immunol.* **2013**, *190*, 3629–3638. [CrossRef] [PubMed]
5. Costa Franco, M.M.; Marim, F.; Guimaraes, E.S.; Assis, N.R.G.; Cerqueira, D.M.; Alves-Silva, J.; Harms, J.; Splitter, G.; Smith, J.; Kanneganti, T.D.; et al. Brucella abortus Triggers a cGAS-Independent STING Pathway To Induce Host Protection That Involves Guanylate-Binding Proteins and Inflammasome Activation. *J. Immunol.* **2018**, *200*, 607–622. [CrossRef] [PubMed]
6. Tupik, J.D.; Coutermarsh-Ott, S.L.; Benton, A.H.; King, K.A.; Kiryluk, H.D.; Caswell, C.C.; Allen, I.C. ASC-Mediated Inflammation and Pyroptosis Attenuates Brucella abortus Pathogenesis Following the Recognition of gDNA. *Pathogens* **2020**, *9*. [CrossRef] [PubMed]
7. Van der Henst, M.; Carlier, E.; De Bolle, X. Intracellular Growth and Cell Cycle Progression are Dependent on (p)ppGpp Synthetase/Hydrolase in Brucella abortus. *Pathogens* **2020**, *9*. [CrossRef] [PubMed]
8. Bialer, M.G.; Sycz, G.; Munoz Gonzalez, F.; Ferrero, M.C.; Baldi, P.C.; Zorreguieta, A. Adhesins of Brucella: Their Roles in the Interaction with the Host. *Pathogens* **2020**, *9*. [CrossRef] [PubMed]
9. Zavattieri, L.; Ferrero, M.C.; Alonso Paiva, I.M.; Sotelo, A.D.; Canellada, A.M.; Baldi, P.C. Brucella abortus Proliferates in Decidualized and Non-Decidualized Human Endometrial Cells Inducing a Proinflammatory Response. *Pathogens* **2020**, *9*. [CrossRef] [PubMed]
10. Gonzalez-Espinoza, G.; Arce-Gorvel, V.; Memet, S.; Gorvel, J.P. Brucella: Reservoirs and Niches in Animals and Humans. *Pathogens* **2021**, *10*. [CrossRef] [PubMed]
11. Santos, R.; Campos, P.C.; Rungue, M.; Rocha, V.; Santos, D.; Mendes, V.; Marinho, F.V.; Martins, F.; Ricci, M.F.; Reis, D.C.D.; et al. The Role of ST2 Receptor in the Regulation of Brucella abortus Oral Infection. *Pathogens* **2020**, *9*. [CrossRef] [PubMed]
12. Arriola Benitez, P.C.; Pesce Viglietti, A.I.; Elizalde, M.M.; Giambartolomei, G.H.; Quarleri, J.F.; Delpino, M.V. Hepatic Stellate Cells and Hepatocytes as Liver Antigen-Presenting Cells during B. abortus Infection. *Pathogens* **2020**, *9*. [CrossRef] [PubMed]
13. Rodriguez, A.M.; Trotta, A.; Melnyczajko, A.P.; Miraglia, M.C.; Kim, K.S.; Delpino, M.V.; Barrionuevo, P.; Giambartolomei, G.H. Brucella abortus-Stimulated Platelets Activate Brain Microvascular Endothelial Cells Increasing Cell Transmigration through the Erk1/2 Pathway. *Pathogens* **2020**, *9*. [CrossRef] [PubMed]
14. Gupta, S.; Mohan, S.; Somani, V.K.; Aggarwal, S.; Bhatnagar, R. Simultaneous Immunization with Omp25 and L7/L12 Provides Protection against Brucellosis in Mice. *Pathogens* **2020**, *9*. [CrossRef] [PubMed]

pathogens MDPI

Article

Simultaneous Immunization with Omp25 and L7/L12 Provides Protection against Brucellosis in Mice

Sonal Gupta [1], Surender Mohan [1], Vikas Kumar Somani [1,2], Somya Aggarwal [1,3] and Rakesh Bhatnagar [1,4,*]

1 Laboratory of Molecular Biology and Genetic Engineering, School of Biotechnology, Jawaharlal Nehru University, New Delhi 110067, India; sonalmole@gmail.com (S.G.); mohan.surender@gmail.com (S.M.); vikass@wustl.edu (V.K.S.); s.aggarwal@wustl.edu (S.A.)
2 Department of Oncology, Washington University School of Medicine, St. Louis, MO 63110, USA
3 Department of Molecular Microbiology, Washington University School of Medicine, St. Louis, MO 63110, USA
4 Banaras Hindu University, Varanasi, Uttar Pradesh 221005, India
* Correspondence: rakeshbhatnagar@jnu.ac.in; Tel.: +91-11-26704079; Fax: +91-11-26717040

Received: 20 January 2020; Accepted: 20 February 2020; Published: 24 February 2020

Abstract: Currently used *Brucella* vaccines, *Brucella abortus* strain 19 and RB51, comprises of live attenuated *Brucella* strains and prevent infection in animals. However, these vaccines pose potential risks to recipient animals such as attenuation reversal and virulence in susceptible hosts on administration. In this context, recombinant subunit vaccines emerge as a safe and competent alternative in combating the disease. In this study, we formulated a divalent recombinant vaccine consisting of Omp25 and L7/L12 of *B. abortus* and evaluated vaccine potential individually as well as in combination. Sera obtained from divalent vaccine (Omp25+L7/L12) immunized mice group exhibited enhanced IgG titers against both components and indicated specificity upon immunoblotting reiterating its authenticity. Further, the IgG1/IgG2a ratio obtained against each antigen predicted a predominant Th2 immune response in the Omp25+L7/L12 immunized mice group. Upon infection with virulent *B. abortus* 544, Omp25+L7/L12 infected mice exhibited superior Log10 protection compared to individual vaccines. Consequently, this study recommends that simultaneous immunization of Omp25 and L7/L12 as a divalent vaccine complements and triggers a Th2 mediated immune response in mice competent of providing protection against brucellosis.

Keywords: Recombinant vaccine; divalent vaccine; brucellosis; Omp25; L7/L12; *Brucella abortus* 544

1. Introduction

Brucellosis, one of the major bacterial zoonoses across the globe, is caused by members of the *Brucella* genus, (*B. abortus*, *B. melitensis*, *B. suis* and *B. canis*). It is still considered one of the seven "neglected zoonoses" worldwide, in spite of a huge public health burden in many countries with low income [1]. The transmission of brucellosis in humans occurs through coincidental exposure to pathogenic bacterium from infected animals or animal products [2–6]. Domestic animals affected by brucellosis are more prone to abortions whereas human brucellosis leads to debilitating symptoms such as recurrent fevers, spondylitis, joint pains, and osteomyelitis [7–9]. The successful vaccines available commercially against brucellosis are Strain 19 and RB51 [10]. Strain 19 is an attenuated *B. abortus* strain with smooth morphology that has the capability to induce antibody responses and protect cattle against brucellosis. *B. abortus* RB51 is a rifampicin-resistant strain with rough morphology which provides protection against infection and abortion [11]. But there are several disadvantages associated with these vaccines such as occurrence of abortions in pregnant cows, restriction on age of vaccination, and reversion of vaccine strain back to pathogenic strain upon administration [12].

Further, antibodies are directed majorly against the lipopolysaccharide O-side chain during natural infection or S19 immunization which causes obstruction during a brucellosis diagnostic test. Therefore, development of an effective and safe vaccine is required.

In this context, recombinant protein-based subunit vaccines have an advantage over traditional live-attenuated vaccines as being protective and safe for human administration [13–15]. Recombinant subunit vaccines rely on specific parts of the pathogenic microorganism such as proteins or capsular polysaccharides containing protective epitopes to result in a protective immune response. Since these vaccines cannot replicate in the host, they are not pathogenic on administration [16]. Numerous intracellular components and surface proteins from bacteria have been researched as potential protective antigens against *Brucella* infection, such as L7/L12 [17], Omp19, Omp31 [18], BLS [19], BP26 [15] and Omp25 [20,21], and some have shown to be effective in providing significant protection against *Brucella* infection [15,17,20,21]. Although recombinant subunit vaccines offer no residual virulence but these require administration of multiple boosters along with providing lower levels of protection [22]. Further, when these *Brucella* protective antigens were administered in combination, the induced immune responses were superior in clearing intracellular *Brucella* as compared to their univalent counterpart [14,15,18,23–27]. Earlier studies have suggested that Omp25 and L7/L12 can serve as efficient protective antigens by inducing strong humoral and cellular immune response [17,20,28]. Outer membrane protein 25 (Omp25) of *Brucella* species has been identified as a potential antigen [20,28] capable of inhibiting TNF-α production [21]. Importantly, *omp25* gene is highly conserved in multiple *Brucella* species, strains, and biovars [8]. In addition, L7/L12 is a conserved protein which is found to be immunogenic and a stimulant of Th1 type $CD4^+$ cellular response in mice [17], making both these antigens relevant as components for divalent vaccine formulation against brucellosis. Further, many protective antigens from *Brucella* have been classified as specific and novel diagnostic target as compared to LPS based conventional tests [29,30].

Aluminum hydroxide (Alum) has been validated as an economical and safe adjuvant by U.S. Food and Drug administration for veterinary and human use. Aluminum salts form a short-term depot at the site of injection and slowly release antigen to the body's immune system [31,32]. In this study, we co-immunized Omp25 and L7/L12 of *Brucella abortus* using alum as an adjuvant. Elicited antibody titers and antibody subtype profile were analyzed when administered as a divalent vaccine candidate in BALB/c mice. Further, the protective efficacy of individual proteins and the divalent vaccine candidate against virulent *B. abortus* 544 challenge were determined.

2. Results

2.1. Expression and Purification of Recombinant Proteins rOmp25 and rL7/L12

The recombinant proteins rOmp25 and rL7/L12 were expressed in *E. coli* C43 cells and *E. coli* BL21 (DE3) cells respectively and purified using Ni-NTA chromatography. Further, the size of the expressed recombinant proteins was verified using 12% SDS-PAGE electrophoresis (Figure 1a). The immunoreactivity of purified proteins was confirmed by immunoblotting using (anti-Omp25+ L7/L12) mice serum signifying that serum from mice immunized with Omp25+L7/L12 consisted of antibodies specifically against its component proteins rOmp25 and rL7/L12 (Figure 1b).

Further, the virulence potential of rOmp25 and rL7/L12 was analyzed using bioinformatics analysis using VirulentPred and VaxiJen. VirulentPred is a tool used for prediction of virulent protein sequences in bacteria based on bi-layer cascade support vector machine (SVM) [33]. The SVM classifiers in this tool were trained and optimized using individual protein sequence features such as their amino acid and dipeptide composition along with position-specific iterated blast (PSI-BLAST). This tool distinguishes virulent proteins from non-virulent bacterial proteins with an accuracy of 81.8% [33]. On the basis of VirulentPred, rOmp25 and rL7/L12 were concluded to be virulent with predicted scores of 1.0411 and 0.2440, respectively (Figure 1c). In addition, we used the VaxiJen tool [34], which uses an alignment-independent approach for prediction of protective antigens. The antigen classification is purely based on the physiochemical properties of proteins without applying sequence alignment [34], and depends on auto cross covariance (ACC) transformation of protein sequences into uniform vectors of amino acid properties. VaxiJen results categorized Omp25 and L7/L12 as vaccine antigens with predicted scores of 0.7506 and 0.6442, respectively, at the threshold value of 0.4 (Figure 1d).

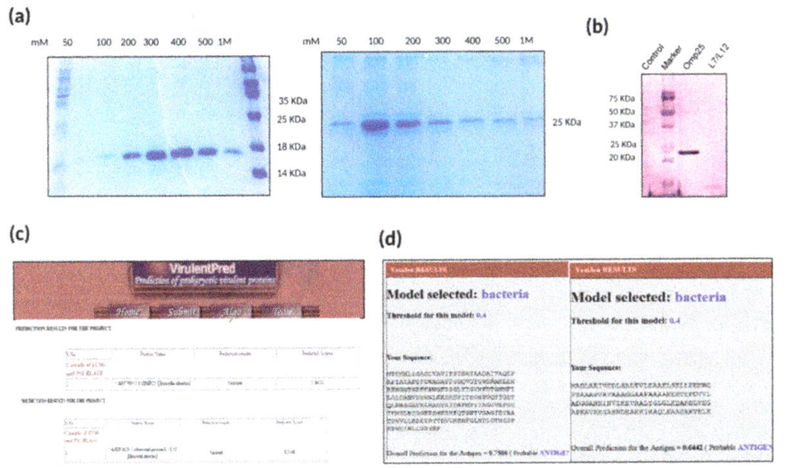

Figure 1. (**a**) Purification of rL7/L12 and rOmp25: SDS-PAGE gel stained with coomassie blue stain showing purification of rL7/L12 and rOmp25 recombinant proteins corresponding to 17 KDa and 25 KDa, respectively. (**b**) Immunoblotting with polyclonal sera of mice immunized with divalent vaccine (Omp25+L7/L12): The reactivity of purified proteins was confirmed by immunoblotting using anti-Omp25+L7/L12 mice serum. Negative control (lane 1; *E. coli* BL21 (DE3) cells with pET28a only), marker (lane 2) Precision Plus Protein™ Dual Color Standards, BIORAD #1610374, rOmp25 (lane 3), and rL7/L12 (lane 4). (**c**) Prediction of virulent proteins in a bacterium using VirulentPred: The sequences for *Brucella abortus* protein, Omp25 and L7/L12, have been submitted as input and their predicted scores have been calculated using VirulentPred software. (**d**) Prediction of vaccine antigens using VaxiJen: The sequences for *Brucella abortus* protein, Omp25 and L7/L12, have been submitted as input and the probability of these proteins being vaccine antigens has been predicted using VaxiJen.

2.2. Determination of IgG Antibody Titre upon Divalent Vaccine Immunization

To assess the levels of IgG antibody titer generated in each of the immunized mice groups, sera was collected at day 28 and 42 post-priming and levels were estimated using Enzyme linked immunosorbent assay (ELISA). Our results revealed that immunization with Omp25+L7/L12 supported a robust anti-L7/L12 IgG response that was detectable at day 28 and remained stable until day 42. At day 28, anti-L7/L12 antibodies were observed to be higher in Omp25+L7/L12 mice compared to L7/L12 only immunized mice ($p < 0.05$; Figure 2a). At day 42, anti-L7/L12 levels were found to be similar in both L7/L12 and Omp25+L7/L12 immunized mice (approximately 6×10^5 in both). Immunization with divalent vaccine elicited a vigorous anti-Omp25 IgG response as well. Antibody levels were observed to be similar in Omp25+L7/L12 and Omp25 only immunized mice at day 28 and day 42 (Figure 2b). Therefore, the antigen alone vaccinated group generated antibodies only against a single antigen, whereas mice immunized with divalent vaccine (Omp25+L7/L12) produced antibodies against both components (rOmp25 and rL7/L12) in a cumulative manner, indicating that co-immunization of two proteins didn't hinder the immune response and supported generation of antibodies against its individual components.

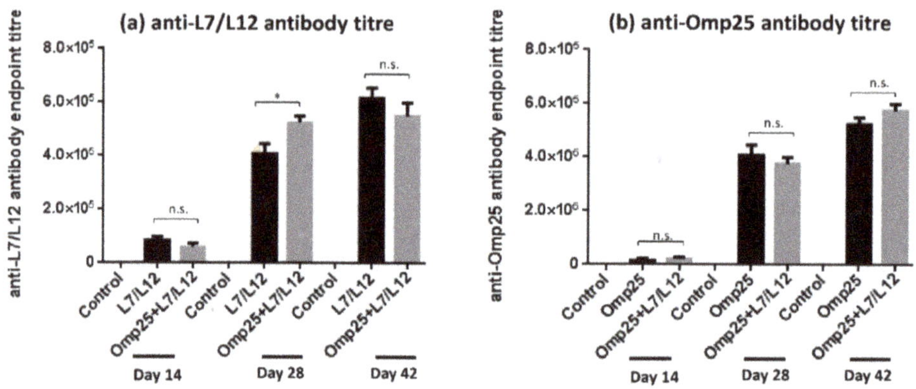

Figure 2. IgG antibody response elicited after immunization with L7/L12, Omp25, and divalent vaccine (Omp25+L7/L12): The mice were immunized with proteins rOmp25, rL7/L12 and rOmp25+rL7/L12 followed by isolation of serum samples from tail veins on day 14, 28, and 42. Estimation of IgG antibody end point titer was done through Enzyme linked immunosorbent assay (ELISA) and data is plotted as (mean ± SD).

2.3. Evaluation of IgG Isotype Levels upon Divalent Vaccine Immunization

In order to predict the Th1/Th2 bias of immune response, the relative IgG isotypes levels (IgG1, IgG2a and IgG2b) were analyzed in Omp25+L7/L12 immunized mice along with mice immunized solely with L7/L12 and Omp25. In the case of anti-L7/L12 antibodies, IgG1 levels were found to be significantly higher than IgG2a levels in the divalent vaccine as well as L7/L12 only immunized mice group, indicating a Th2 biased immune response in both (Figure 3a). Similarly, in the case of anti-Omp25 antibody levels, IgG1 levels were found to be higher than IgG2a (IgG1/IgG2a = 1.86), suggesting a Th2 biased immune response in divalent vaccine immunized mice. Interestingly, levels of IgG2a and IgG2b antibodies in Omp25+L7/L12 immunized mice were noteworthy (Figure 3b), predicting an elicitation of Th1 immune response in divalent vaccine immunized mice as well.

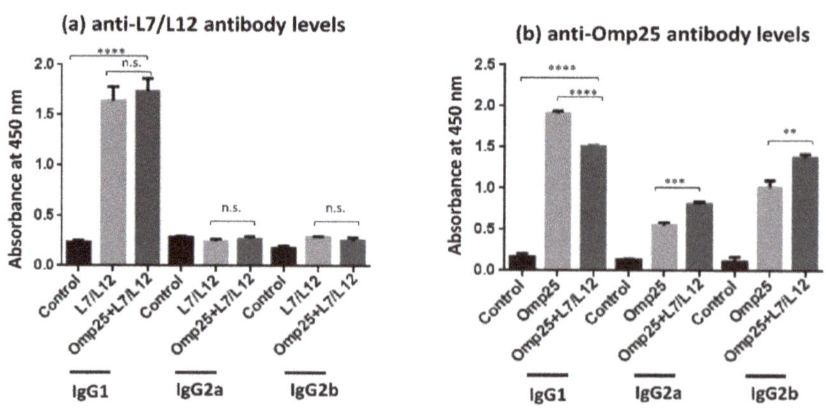

Figure 3. IgG antibody isotypes (IgG1, IgG2a, and IgG2b) elicited after immunization with rL7/L12, rOmp25, and divalent vaccine (Omp25+L7/L12): The recombinant *B. abortus* proteins rL7/L12, rOmp25, and Omp25+ L7/L12 were immunized in mice and isolation of serum samples was done from tail veins on day 42. Estimation of IgG isotype levels in serum of immunized mice was done through ELISA. The antibodies used for ELISA were horseradish peroxidase (HRP) conjugated anti-mouse IgG1, IgG2a, and IgG2b antibodies and data is plotted as (mean (OD_{450} ± SD)).

2.4. Evaluation of Protective Efficacy Conferred by Divalent Vaccine Candidate against B. abortus 544 Challenge

The protective efficacy of the Omp25+L7/L12 vaccine candidate along with groups immunized solely with Omp25 and L7/L12 was analyzed against *B. abortus* 544 infection. Two weeks after last immunization, the immunized mice were challenged with virulent *B. abortus* 544 through intraperitoneal route. The mice were sacrificed four weeks post-infection, and bacterial colony forming units (CFU) were determined. As shown in Table 1, the level of $\log_{10}$ CFU per spleen at 28 days post-challenge with *B. abortus* 544 was (4.820 ± 0.18) in Omp25+L7/L12 immunized mice. Consecutively, $\log_{10}$ protection conferred by the Omp25+L7/L12 group was 1.98 at 28 day post-challenge as compared to (PBS + alum) immunized mice indicating that the Omp25+L7/L12 vaccine candidate was effective at eliminating pathogenic *B. abortus* 544 from a mice model. Mice immunized with Omp25 and L7/L12 alone exhibited $\log_{10}$ units of protection as 1.46 and 1.75, respectively at 28 days post-challenge with *B. abortus* 544 as compared to (PBS + alum) immunized mice. Overall, upon analyzing the levels of protection of the divalent vaccine candidate against *B. abortus* 544 challenge, it was found that Omp25+L7/L12 immunized mice exhibited efficacious $\log_{10}$ units of protection against *B. abortus* 544 challenge along with its individual components, however S19 exhibited the maximum.

Table 1. Bacterial proliferation in the spleen of mice immunized with rOmp25, rL7/L12, divalent vaccine candidate (Omp25+L7/L12) and control, using alum as adjuvant. The mice were infected with *B. abortus* 544 through intraperitoneal route and the splenic bacterial load was determined by plating dilutions of the splenocytes suspension on the tryptic soya agar plates followed by incubation at 37 °C in the presence of 5% CO_2 for 48 h. Data is represented as mean ± S.D.

S. No	Vaccine Candidate	Log_{10} Spleen Bacillary Load at Day 28 Post-Challenge (Log_{10} CFU)	Log_{10} Units of Protection at Day 28 Post-Challenge
1.	rOmp25	5.338 ± 0.75	1.46
2.	rL7/L12	5.05 ± 0.27	1.75
3.	Omp25+L7/L12	4.820 ± 0.18	1.98
4.	PBS	6.80 ± 0.58	–
5.	*B. abortus* S19	4.21 ± 0.31	2.59

3. Discussion

Brucellosis is still a major public health concern and endemic in many countries, mainly in the Mediterranean region, eastern and western Africa, and parts of South and Central America. There is a substantial requirement to control and eradicate this disease caused by the *Brucella* genus [4,5,7,35,36]. The current vaccines in use, Strain 19 and RB51 despite being popular are still far from ideal [12,37]. They prevent infection in animal but offer potential risks such as attenuation reversal and virulence in susceptible hosts. In this context, subunit vaccines are better options as compared to live attenuated vaccines, since they are safe and do not revert back to pathogenic strain upon administration [13–15]. Further, recombinant subunit vaccines protect against a given pathogen by activating humoral and cellular arms of immunity based upon a specific antigen along with used adjuvant, which makes them competent and useful in the vaccine field. There are certain proteins in *Brucella* species which can provide significant protection against the disease and are conserved throughout, such as L7/L12 [17], Omp19, Omp31 [18], BLS [19], BP26 [15], and Omp25 [20,21]. Among these, *omp25* and *l7/l12* genes in *Brucella* species encode for Omp25 and L7/L12 immunodominant proteins respectively, and have the potential to stimulate a strong humoral immune response along with providing protection against *Brucella* infections in mice models [15,17,20,28]. Earlier studies have suggested that outer membrane proteins from *Brucella* such as Omp25 [38], Omp10 [39], Omp19 [39], BP26 [40], Omp28 [41,42], and Omp31 [30] can distinguish between *Brucella*-infected animals and non-infected ones in an efficient and accurate way, withdrawing the false positive results in the field due to cross-reacting antibodies.

Further multivalent subunit vaccine formulations possess the capability to generate a wide range of immunogens that may result in better protection than their univalent counterpart [24–26]. In this study, we evaluated humoral immune response and protective efficacy of a divalent vaccine candidate consisting of rOmp25 and rL7/L12 against *Brucella* infection in mice. The foremost point to be explored in this study was whether two components combined together in a divalent vaccine have the capability to show a synergistic response and promote a heightened immune response, or if some kind of competitive inhibition occurs among them? Aluminum hydroxide (Alum) is beneficial since it is inexpensive and has been certified as the safest adjuvant for use by the United States Food and Drug Administration [31,32,43]. Alum creates a depot effect at the site of injection, resulting in a slow release of adsorbed antigens and an elevation in the immune response. The intraperitoneal route of administration was chosen because it helps to quickly absorb antigens into the vasculature, which leads to a rise of antigen drainage into the spleen and activation of immune cells circulating in the lymph nodes [44].

The results in this study exhibited that humoral immune response was elevated in mice immunized with the divalent vaccine Omp25+L7/L12 as compared to the control group (PBS + alum). The divalent vaccine (Omp25+L7/L12) was found immunogenic with high IgG levels against both of its components, rOmp25 and rL7/L12 (Figure 2), depicting that divalent vaccine has the potential to exhibit synergy among its individual components and elevate the immune response against virulent *Brucella abortus* 544 challenge. During bacterial infection, Th2-mediated immune response is characterized by synthesis and increase in the level of IgG1 antibodies [45,46] whereas Th1 immune response is represented by levels of IgG2a antibody along with IFN-γ cytokine levels. The IgG subclass (IgG1, IgG2a and IgG2b) detection in Omp25+L7/L12 immunized mice exhibited that IgG1 levels were significantly higher as compared to IgG2a levels, predicting a more prominent Th2 immune response in case of anti-Omp25 and anti-L7/L12 antibodies (Figure 3). Although individual vaccinated group generated antibodies specifically to single antigen immunized whereas divalent vaccine candidate resulted in generation of antibodies against both antigens, rOmp25 and rL7/L12. Further, immunization with alum as an adjuvant has also been suggested to enhance antibody response in mice [43,47]. It helps in enhancement of antigen uptake and presentation to antigen-presenting cells, which results in promotion of Th2 immune responses [47,48]. Therefore, it is possible that alum has helped to increase antibody titer and elevate the humoral immune response in divalent vaccine immunized mice [49]. The analysis of protective efficacy in different mice groups after infection with virulent *Brucella abortus* 544 showed that Omp25+L7/L12 immunized mice exhibited a significant increase in $\log_{10}$ protection (1.98) as compared to the control (i.e. alum immunized mice) at 28 days after challenge (p value < 0.001; Table 1). This specifies that Omp25+L7/L12 immunized mice were capable of eliminating pathogenic *B. abortus* 544 compared to the control group. On comparing $\log_{10}$ units of protection at 28 days after challenge in individual protein immunized mice, rOmp25 (1.46) and rL7/12 (1.75) with alum as the adjuvant, it was observed that Omp25+L7/L12 immunized mice showed a superior level of protection against *B. abortus* 544 infection, however S19 exhibited the maximum.

It is noteworthy that although *B. abortus* recombinant subunit vaccines show very promising results in mice models, the immune responses recognized in mice models may not reflect the protection achieved in natural hosts such as cattle after immunization [11]. Therefore, further studies determining protective efficacy in other animal models such as rats, guinea pigs, and monkeys are also encouraged before proceeding towards cattle administration [50]. Recombinant vaccines also need multiple booster administrations along with adjuvants and a combination of several antigens, which makes them economically unsuitable for cattle immunization [51]. Hence, there is a need to decrease the production cost, search for effective and affordable adjuvants, and reduce the expense of recombinant protein purification in order to make these vaccines economical for mass administration.

In a nutshell, this preliminary study shows that the combination of rOmp25 and rL7/L12 elicited steady immune responses against both antigens in mice. Further, when mice were immunized with the Omp25+L7/L12 vaccine candidate, a significant reduction in *B. abortus* 544 load in mice spleens was observed, implying the use of divalent vaccine (Omp25+L7/L12) as an improved vaccine candidate against brucellosis. Nevertheless, this study illustrates the potential of a divalent vaccine in providing host immunity and protection against *B. abortus* challenge, suggesting the use of a divalent recombinant vaccine candidate as an advanced approach in the future against brucellosis.

4. Materials and Methods

4.1. Plasmids and Bacterial Strains

E. coli DH5α was used for propagation of recombinant plasmids. *E. coli* BL21 (DE3) and C43 strains were used for expression of rL7/L12 and rOmp25 proteins, respectively. *E. coli* strains were cultured using Luria–Bertani (LB) medium. Kanamycin was added to the medium at a final concentration of 50 μg/mL. *B. abortus* 544 and S19 strains were obtained from the Indian Veterinary Research Institute, Bareilly, India. *Brucella abortus* 544 was cultured in tryptic soy medium. Experiments involving *B. abortus* 544 and S19 strains were performed in a biosafety Level 3 laboratory at Jawaharlal Nehru University (JNU), Delhi, India.

4.2. Expression and Purification of Recombinant Proteins

For formulation of the divalent vaccine, Omp25 and L7/L12 antigens of *Brucella abortus* were PCR amplified using gene specific primers and cloned in pET28(a) vector (Table 2). The expression of proteins was done in *E. coli*. To purify rOmp25, recombinants were grown in terrific broth until OD_{600} ~ 0.5–0.6, and then induction was done using 1 mM IPTG for 5 h at 37 °C. Further purification of rOmp25 was done from the insoluble inclusion bodies fraction, using the urea-denaturing method and on-column refolding [20]. To purify rL7/L12, recombinants were grown in LB medium containing kanamycin to OD_{600} ~ 0.7–0.8 followed by induction using 1 mM IPTG for 5 h at 37 °C. Both the proteins were affinity purified using immobilized nickel-nitrilotriacetic acid (Ni-NTA) agarose columns equilibrated in PB buffer (100 mM potassium phosphate buffer, pH 8.0) and eluted using 100–500 mM imidazole in PB. Purified proteins were analyzed by SDS-PAGE for content and purity. The dialysis of purified proteins was done against phosphate buffer saline (pH 7.4).

Table 2. Description of strains used in this study.

S. No.	Protein Name	Strain Used for Purification of Protein	Reference
1.	rOmp25	*omp25* gene was cloned in pET28a at BamHI and SalI sites and expressed in *E. coli* C43 cells.	Goel et al. 2012 [20]
2.	rL7/L12	L7/L12 ribosomal gene was cloned in pET28a at NcoI and XhoI sites and expressed in *E. coli* BL21(DE3) cells.	Singh et al. 2015 [1]

4.3. Immunoblotting

For immunoblotting, the recombinant proteins were resolved by 12% SDS-PAGE followed by electroblotting onto nitrocellulose membrane. The membrane was further blocked using 3% BSA followed by incubation with anti-Omp25+L7/L12 antibody (1:5000 dilution, raised in mice) for 1 h. After subsequent washing, binding specificity was checked using AP-conjugated goat anti-mouse IgG antibody (Catalog no. sc-2047, Santa Cruz Biotechnology, USA) [52–54].

4.4. Immunization of Respective Proteins in Mice

Four to six week old female BALB/c mice (inbred) were obtained from the National Centre for Laboratory Animal Sciences, Hyderabad, India. Recommendations from the Institutional Animal Ethics and Biosafety Committee were regularly followed during mouse experiments. In brief, mice were caged under sterile conditions in micro-isolators, fed with pathogen-free food and water *ad libitum* during consecutive immunizations. Once infected with *B. abortus* 544, mice were maintained at the BSL-3 animal facility of JNU for evaluation of protective efficacy.

For immunization of rL7/L12 and rOmp25, the optimized dose of each antigen was considered as mentioned in earlier reports [1,20]. Briefly, mice were grouped and immunized through the intraperitoneal route, either with Omp25 (30 µg) or L7/L12 (40 µg) alone or in combination as a divalent vaccine candidate with alum as an adjuvant. Two boosters were administered at regular intervals of 2 weeks, and 1X PBS with alum and *B. abortus* S19 immunized mice groups were taken as controls. For prime immunization and subsequent booster immunization, 100 µl emulsion of the required antigen and alum in 1X PBS was injected in each mouse. The blood was collected from each mouse on day 0, 14, 28, and 42 from tail veins and sera was extracted through centrifugation at 15,600 *g* for 20 min, followed by storage at −80 °C for further analysis.

4.5. Elucidation of End-Point Antibody Titer

An enzyme-linked immunosorbent assay (ELISA) was used to analyze serum antibody titer. In brief, 96-well microtiter plates (NuncMaxiSorp) were coated overnight with 500 ng/well of capture antigen (rOmp25 or rL7/L12) in PBS at 4 °C. The plates were washed three times using PBST (PBS with 0.1% tween 20) followed by blocking using 2% BSA in PBS for 2 h at 37 °C. The antibody titer in the sera of respective antigen immunized mice along with the divalent vaccine immunized mouse group was assessed by priming dilutions of the same, in triplicates, at 37 °C for 1 h. Washing of the plates was done using PBST followed by addition of horseradish peroxidase (HRP)-conjugated anti-mouse secondary antibodies (Catalog no. sc-2005, Santa Cruz Biotechnology, USA) at 1:10,000 dilution for 1 h at 37 °C [53,54]. The plates were further incubated with OptEIA TMB substrate (BD Biosciences, USA) for calorimetric assay and the reaction was stopped using 1N HCl. Absorbance of the plate was measured at 450 nm through Tecan's Sunrise absorbance microplate reader. End point titer was evaluated as the reciprocal of highest dilution giving absorbance greater than the threshold value. Threshold value was calculated as the mean of absorbance plus three times standard deviation of 1:1000 dilution of the control group (PBS + alum).

4.6. Analysis of IgG Isotypes in Immunized Mice

The IgG isotypes (IgG1, IgG2a and IgG2b) were detected in immunized mice using ELISA as described above. For secondary antibodies, anti-mouse IgG1-HRP (Catalog no. sc-2060), anti-mouse IgG2a-HRP (Catalog no. sc-2061) and anti-mouse IgG2b-HRP conjugated antibodies (Catalog no. sc-2062) (raised in goat; Santa Cruz Biotechnology, USA) were used and absorbance at 450 nm was measured [53].

4.7. Evaluation of Protective Efficacy of Vaccine Candidate

Two weeks after the final booster immunization (day 42), mice groups immunized with PBS, rOmp25, rL7/L12, and divalent vaccine candidate (rOmp25+rL7/L12) were challenged with 2×10^5 cells of *B. abortus* 544 through the intraperitoneal route. *B.abortus* S19 was injected on day 0 in respective group, and challenge was done after 21 days with 2×10^5 cells of virulent *B. abortus* 544. After 4 weeks of infection, mice from each group were euthanized through cervical dislocation. Their spleen was extracted under sterile conditions and finally homogenized in PBS using probe homogenizer. For CFU count, various dilutions of the spleen homogenate were prepared and plated on tryptic soya agar followed by incubation at 37 °C for 48 h in the presence of 5% CO_2. Total splenic load was calculated and represented as Log_{10} CFU mean ± standard deviation (SD). Log_{10} units of protection were determined by calculating the difference between the log_{10} CFU of PBS injected group (control) and vaccinated group.

4.8. Statistical Analysis

The results are represented as mean ± SD and are reported as data of three different sets of experiments. The statistical significance in antibody titer was calculated using two-tailed Student's t-test. (* represents $P < 0.05$; ** represents $P < 0.01$; *** represents $P < 0.001$, **** represents $P < 0.0001$).

Ethical statement: All mice experiments were performed while abiding by the rules of Institutional Animal Ethics Committee (IAEC), Jawaharlal Nehru University, New Delhi, India guidelines. All experiments involving virulent *Brucella abortus* 544 and *Brucella abortus* S19 strain have been performed in Biosafety level-3 (BSL-3) facility.

Author Contributions: Each author has made valuable contributions to this work. Conceptualization, S.G. and R.B.; methodology, S.G., RB; validation, S.G., S.M., V.K.S., S.A. and R.B.; formal analysis, S.G., V.K.S., S.A.; investigation, R.B.; resources, R.B.; data curation, S.G.; writing—original draft preparation, S.G.; writing—review and editing, S.G., S.M., V.K.S., S.A., R.B.; visualization, S.G., S.M., V.K.S. and S.A.; supervision, R.B.; project administration, R.B.; funding acquisition, R.B. All authors have read and agreed to the published version of the manuscript.

Funding: We would like to thank the Department of Biotechnology, Government of India for providing us with financial support and specifically, we acknowledge members of "Network Project on Brucellosis" for their valuable guidance and support. S. Gupta is a Senior Research Fellowship recipient from the Council of Scientific & Industrial Research (*CSIR*), Government of India (Grant no. 09/263(1085)2015-EMR-1).

Conflicts of Interest: The authors declared no conflict of interest.

References

1. Singh, D.; Goel, D.; Bhatnagar, R. Recombinant L7/L12 protein entrapping PLGA (poly lactide-*co*-glycolide) micro particles protect BALB/c mice against the virulent B. abortus 544 infection. *Vaccine* **2015**, *33*, 2786–2792. [CrossRef] [PubMed]
2. Moreno, E.; Stackebrandt, E.; Dorsch, M.; Wolters, J.; Busch, M.; Mayer, H. Brucella abortus 16S rRNA and lipid A reveal a phylogenetic relationship with members of the alpha-2 subdivision of the class Proteobacteria. *J. Bacteriol.* **1990**, *172*, 3569–3576. [CrossRef] [PubMed]
3. Renukaradhya, G.J.; Isloor, S.; Rajasekhar, M. Epidemiology, zoonotic aspects, vaccination and control/eradication of brucellosis in India. *Vet. Microbiol.* **2002**, *90*, 183–195. [CrossRef]
4. Godfroid, J.; Cloeckaert, A.; Liautard, J.P.; Kohler, S.; Fretin, D.; Walravens, K.; Garin-Bastuji, B.; Letesson, J.J. From the discovery of the Malta fever's agent to the discovery of a marine mammal reservoir, brucellosis has continuously been a re-emerging zoonosis. *Vet. Res.* **2005**, *36*, 313–326. [CrossRef] [PubMed]
5. Tan, S.Y.; Davis, C. David Bruce (1855–1931): Discoverer of brucellosis. *Singap. Med. J.* **2011**, *52*, 138–139.
6. Babaoglu, U.T.; Ogutcu, H.; Demir, G.; Sanli, D.; Babaoglu, A.B.; Oymak, S. Prevalence of Brucella in raw milk: An example from Turkey. *Niger. J. Clin. Pract.* **2018**, *21*, 907–911.
7. Glynn, M.K.; Lynn, T.V. Brucellosis. *J. Am. Vet. Med. Assoc.* **2008**, *233*, 900–908. [CrossRef]
8. Cloeckaert, A.; Verger, J.M.; Grayon, M.; Grepinet, O. Restriction site polymorphism of the genes encoding the major 25 kDa and 36 kDa outer-membrane proteins of Brucella. *Microbiology* **1995**, *141*, 2111–2121. [CrossRef]
9. Pappas, G.; Bosilkovski, M.; Akritidis, N.; Mastora, M.; Krteva, L.; Tsianos, E. Brucellosis and the respiratory system. *Clin. Infect. Dis.* **2003**, *37*, e95–e99. [CrossRef]
10. Olsen, S.C.; Stoffregen, W.S. Essential role of vaccines in brucellosis control and eradication programs for livestock. *Expert Rev. Vaccines* **2005**, *4*, 915–928. [CrossRef]
11. Dorneles, E.M.; Sriranganathan, N.; Lage, A.P. Recent advances in Brucella abortus vaccines. *Vet. Res.* **2015**, *46*, 76. [CrossRef] [PubMed]
12. Nicoletti, P. Prevalence and persistence of Brucella abortus strain 19 infections and prevalence of other biotypes in vaccinated adult dairy cattle. *J. Am. Vet. Med. Assoc.* **1981**, *178*, 143–145. [PubMed]
13. Nascimento, I.P.; Leite, L.C.C. Recombinant vaccines and the development of new vaccine strategies. *Braz. J. Med Biol. Res. Rev. Bras. De Pesqui. Med. E Biol.* **2012**, *45*, 1102–1111. [CrossRef] [PubMed]

14. Hop, H.T.; Arayan, L.T.; Huy, T.X.N.; Reyes, A.W.B.; Min, W.; Lee, H.J.; Park, S.J.; Chang, H.H.; Kim, S. Immunization of BALB/c mice with a combination of four recombinant Brucella abortus proteins, AspC, Dps, InpB and Ndk, confers a marked protection against a virulent strain of Brucella abortus. *Vaccine* **2018**, *36*, 3027–3033. [CrossRef]
15. Yang, X.; Walters, N.; Robison, A.; Trunkle, T.; Pascual, D.W. Nasal immunization with recombinant Brucella melitensis bp26 and trigger factor with cholera toxin reduces B. melitensis colonization. *Vaccine* **2007**, *25*, 2261–2268. [CrossRef]
16. Hansson, M.; Nygren, P.A.; Stahl, S. Design and production of recombinant subunit vaccines. *Biotechnol. Appl. Biochem.* **2000**, *32*, 95–107. [CrossRef]
17. Oliveira, S.C.; Splitter, G.A. Immunization of mice with recombinant L7/L12 ribosomal protein confers protection against Brucella abortus infection. *Vaccine* **1996**, *14*, 959–962. [CrossRef]
18. Cassataro, J.; Estein, S.M.; Pasquevich, K.A.; Velikovsky, C.A.; de la Barrera, S.; Bowden, R.; Fossati, C.A.; Giambartolomei, G.H. Vaccination with the recombinant Brucella outer membrane protein 31 or a derived 27-amino-acid synthetic peptide elicits a CD^{4+} T helper 1 response that protects against Brucella melitensis infection. *Infect. Immun.* **2005**, *73*, 8079–8088. [CrossRef]
19. Velikovsky, C.A.; Goldbaum, F.A.; Cassataro, J.; Estein, S.; Bowden, R.A.; Bruno, L.; Fossati, C.A.; Giambartolomei, G.H. Brucella lumazine synthase elicits a mixed Th_1-Th_2 immune response and reduces infection in mice challenged with Brucella abortus 544 independently of the adjuvant formulation used. *Infect. Immun.* **2003**, *71*, 5750–5755. [CrossRef]
20. Goel, D.; Bhatnagar, R. Intradermal immunization with outer membrane protein 25 protects Balb/c mice from virulent B. abortus 544. *Mol. Immunol.* **2012**, *51*, 159–168. [CrossRef]
21. Paul, S.; Peddayelachagiri, B.V.; Nagaraj, S.; Kingston, J.J.; Batra, H.V. Recombinant outer membrane protein 25c from Brucella abortus induces Th_1 and Th_2 mediated protection against Brucella abortus infection in mouse model. *Mol. Immunol.* **2018**, *99*, 9–18. [CrossRef] [PubMed]
22. Lalsiamthara, J.; Lee, J.H. Development and trial of vaccines against Brucella. *J. Vet. Sci.* **2017**, *18*, 281–290. [CrossRef] [PubMed]
23. Al-Mariri, A.; Tibor, A.; Mertens, P.; De Bolle, X.; Michel, P.; Godefroid, J.; Walravens, K.; Letesson, J.J. Protection of BALB/c mice against Brucella abortus 544 challenge by vaccination with bacterioferritin or P39 recombinant proteins with CpG oligodeoxynucleotides as adjuvant. *Infect. Immun.* **2001**, *69*, 4816–4822. [CrossRef] [PubMed]
24. Delpino, M.V.; Estein, S.M.; Fossati, C.A.; Baldi, P.C.; Cassataro, J. Vaccination with Brucella recombinant DnaK and SurA proteins induces protection against Brucella abortus infection in BALB/c mice. *Vaccine* **2007**, *25*, 6721–6729. [CrossRef]
25. Ghasemi, A.; Jeddi-Tehrani, M.; Mautner, J.; Salari, M.H.; Zarnani, A.H. Simultaneous immunization of mice with Omp31 and TF provides protection against Brucella melitensis infection. *Vaccine* **2015**, *33*, 5532–5538. [CrossRef]
26. Tabatabai, L.B.; Pugh, G.W. Modulation of immune responses in Balb/c mice vaccinated with Brucella abortus Cu Zn superoxide dismutase synthetic peptide vaccine. *Vaccine* **1994**, *12*, 919–924. [CrossRef]
27. Tadepalli, G.; Singh, A.K.; Balakrishna, K.; Murali, H.S.; Batra, H.V. Immunogenicity and protective efficacy of Brucella abortus recombinant protein cocktail (rOmp19+rP39) against B. abortus 544 and B. melitensis 16M infection in murine model. *Mol. Immunol.* **2016**, *71*, 34–41. [CrossRef]
28. Goel, D.; Rajendran, V.; Ghosh, P.C.; Bhatnagar, R. Cell mediated immune response after challenge in Omp25 liposome immunized mice contributes to protection against virulent Brucella abortus 544. *Vaccine* **2013**, *31*, 1231–1237. [CrossRef]
29. Vatankhah, M.; Beheshti, N.; Mirkalantari, S.; Khoramabadi, N.; Aghababa, H.; Mahdavi, M. Recombinant Omp2b antigen-based ELISA is an efficient tool for specific serodiagnosis of animal brucellosis. *Braz. J. Microbiol. Publ. Braz. Soc. Microbiol.* **2019**, *50*, 979–984. [CrossRef]
30. Ahmed, I.M.; Khairani-Bejo, S.; Hassan, L.; Bahaman, A.R.; Omar, A.R. Serological diagnostic potential of recombinant outer membrane proteins (rOMPs) from Brucella melitensis in mouse model using indirect enzyme-linked immunosorbent assay. *BMC Vet. Res.* **2015**, *11*, 275. [CrossRef]
31. Gupta, R.K.; Rost, B.E.; Relyveld, E.; Siber, G.R. Adjuvant properties of aluminum and calcium compounds. *Pharm. Biotechnol.* **1995**, *6*, 229–248. [CrossRef] [PubMed]

32. Sivakumar, S.M.; Safhi, M.M.; Kannadasan, M.; Sukumaran, N. Vaccine adjuvants—Current status and prospects on controlled release adjuvancity. *Saudi. Pharm. J.* **2011**, *19*, 197–206. [CrossRef] [PubMed]
33. Garg, A.; Gupta, D. VirulentPred: A SVM based prediction method for virulent proteins in bacterial pathogens. *BMC Bioinform.* **2008**, *9*, 62. [CrossRef] [PubMed]
34. Doytchinova, I.A.; Flower, D.R. VaxiJen: A server for prediction of protective antigens, tumour antigens and subunit vaccines. *BMC Bioinform.* **2007**, *8*, 4. [CrossRef] [PubMed]
35. Fugier, E.; Pappas, G.; Gorvel, J.P. Virulence factors in brucellosis: Implications for aetiopathogenesis and treatment. *Expert Rev. Mol. Med.* **2007**, *9*, 1–10. [CrossRef]
36. Martirosyan, A.; Moreno, E.; Gorvel, J.P. An evolutionary strategy for a stealthy intracellular Brucella pathogen. *Immunol. Rev.* **2011**, *240*, 211–234. [CrossRef]
37. Moriyon, I.; Grillo, M.J.; Monreal, D.; Gonzalez, D.; Marin, C.; Lopez-Goni, I.; Mainar-Jaime, R.C.; Moreno, E.; Blasco, J.M. Rough vaccines in animal brucellosis: Structural and genetic basis and present status. *Vet. Res.* **2004**, *35*, 1–38. [CrossRef]
38. Cloeckaert, A.; Zygmunt, M.S.; Bezard, G.; Dubray, G. Purification and antigenic analysis of the major 25-kilodalton outer membrane protein of Brucella abortus. *Res. Microbiol.* **1996**, *147*, 225–235. [CrossRef]
39. Simborio, H.L.; Lee, J.J.; Bernardo Reyes, A.W.; Hop, H.T.; Arayan, L.T.; Min, W.; Lee, H.J.; Yoo, H.S.; Kim, S. Evaluation of the combined use of the recombinant Brucella abortus Omp10, Omp19 and Omp28 proteins for the clinical diagnosis of bovine brucellosis. *Microb. Pathog.* **2015**, *83–84*, 41–46. [CrossRef]
40. Cloeckaert, A.; Baucheron, S.; Vizcaino, N.; Zygmunt, M.S. Use of recombinant BP26 protein in serological diagnosis of Brucella melitensis infection in sheep. *Clin. Diagn. Lab. Immunol.* **2001**, *8*, 772–775. [CrossRef]
41. Salih-Alj Debbarh, H.; Cloeckaert, A.; Bezard, G.; Dubray, G.; Zygmunt, M.S. Enzyme-linked immunosorbent assay with partially purified cytosoluble 28-kilodalton protein for serological differentiation between Brucella melitensis-infected and B. melitensis Rev.1-vaccinated sheep. *Clin. Diagn. Lab. Immunol.* **1996**, *3*, 305–308. [CrossRef] [PubMed]
42. Chaudhuri, P.; Prasad, R.; Kumar, V.; Gangaplara, A. Recombinant OMP28 antigen-based indirect ELISA for serodiagnosis of bovine brucellosis. *Mol. Cell. Probes* **2010**, *24*, 142–145. [CrossRef] [PubMed]
43. Awate, S.; Babiuk, L.A.; Mutwiri, G. Mechanisms of action of adjuvants. *Front. Immunol.* **2013**, *4*, 114. [CrossRef] [PubMed]
44. Olin, T.; Saldeen, T.J.C.R. The lymphatic pathways from the peritoneal cavity: A lymphangiographic study in the rat. *Cancer Res.* **1964**, *24*, 1700–1711. [PubMed]
45. Maecker, H.T.; Do, M.S.; Levy, S. CD81 on B cells promotes interleukin 4 secretion and antibody production during T helper type 2 immune responses. *Proc. Natl. Acad. Sci. USA* **1998**, *95*, 2458–2462. [CrossRef] [PubMed]
46. Pasquevich, K.A.; Estein, S.M.; Garcia Samartino, C.; Zwerdling, A.; Coria, L.M.; Barrionuevo, P.; Fossati, C.A.; Giambartolomei, G.H.; Cassataro, J. Immunization with recombinant Brucella species outer membrane protein Omp16 or Omp19 in adjuvant induces specific CD^{4+} and CD^{8+} T cells as well as systemic and oral protection against Brucella abortus infection. *Infect. Immun.* **2009**, *77*, 436–445. [CrossRef]
47. Ghimire, T.R. The mechanisms of action of vaccines containing aluminum adjuvants: An in vitro vs in vivo paradigm. *SpringerPlus* **2015**, *4*, 181. [CrossRef]
48. Mannhalter, J.W.; Neychev, H.O.; Zlabinger, G.J.; Ahmad, R.; Eibl, M.M. Modulation of the human immune response by the non-toxic and non-pyrogenic adjuvant aluminium hydroxide: Effect on antigen uptake and antigen presentation. *Clin. Exp. Immunol.* **1985**, *61*, 143–151.
49. Audibert, F.M.; Lise, L.D. Adjuvants: Current status, clinical perspectives and future prospects. *Immunol. Today* **1993**, *14*, 281–284. [CrossRef]
50. Silva, T.M.; Costa, E.A.; Paixao, T.A.; Tsolis, R.M.; Santos, R.L. Laboratory animal models for brucellosis research. *J. Biomed. Biotechnol.* **2011**, *2011*, 518323. [CrossRef]
51. Schurig, G.G.; Sriranganathan, N.; Corbel, M.J. Brucellosis vaccines: Past, present and future. *Vet. Microbiol.* **2002**, *90*, 479–496. [CrossRef]
52. Ghosh, S.; Sulistyoningrum, D.C.; Glier, M.B.; Verchere, C.B.; Devlin, A.M. Altered glutathione homeostasis in heart augments cardiac lipotoxicity associated with diet-induced obesity in mice. *J. Biol. Chem.* **2011**, *286*, 42483–42493. [CrossRef] [PubMed]

53. Gupta, S.; Singh, D.; Gupta, M.; Bhatnagar, R. A combined subunit vaccine comprising BP26, Omp25 and L7/L12 against brucellosis. *Pathog. Dis.* **2019**, *77*. [CrossRef] [PubMed]
54. Aggarwal, S.; Somani, V.K.; Gupta, S.; Garg, R.; Bhatnagar, R. Development of a novel multiepitope chimeric vaccine against anthrax. *Med. Microbiol. Immunol.* **2019**, *208*, 185–195. [CrossRef]

pathogens

MDPI

Article

The Role of ST2 Receptor in the Regulation of *Brucella abortus* Oral Infection

Raiany Santos [1], Priscila C. Campos [2], Marcella Rungue [2], Victor Rocha [2], David Santos [2], Viviani Mendes [2], Fabio V. Marinho [2], Flaviano Martins [3], Mayra F. Ricci [4], Diego C. dos Reis [4], Geovanni D. Cassali [4], José Carlos Alves-Filho [5], Angelica T. Vieira [2,†] and Sergio C. Oliveira [2,*,†]

1 Department of Genetics, Institute of Biological Sciences, Federal University of Minas Gerais—Belo Horizonte, Minas Gerais 31270-901, Brazil; raianyas@yahoo.com.br

2 Department of Biochemistry and Immunology, Institute of Biological Sciences, Federal University of Minas Gerais—Belo Horizonte, Minas Gerais 31270-901, Brazil; pccampos78@gmail.com (P.C.C.); marcellarungue@hotmail.com (M.R.); victormelobio@gmail.com (V.R.); davidmartinsst@gmail.com (D.S.); mendesviviani.a@gmail.com (V.M.); fabiovitarelli@yahoo.com.br (F.V.M.); angelicathomaz@gmail.com (A.T.V.)

3 Department of Microbiology, Institute of Biological Sciences, Federal University of Minas Gerais—Belo Horizonte, Minas Gerais 31270-901, Brazil; flaviano@icb.ufmg.br

4 Department of General Pathology, Institute of Biological Sciences, Federal University of Minas Gerais—Belo Horizonte, Minas Gerais 31270-901, Brazil; riccimayra@gmail.com (M.F.R.); diegobiomed.reis@gmail.com (D.C.d.R.); geovanni.cassali@gmail.com (G.D.C.)

5 Department of Pharmacology, Ribeirao Preto Medical School, University of Sao Paulo, Ribeirao Preto 14049-900, Brazil; jcafilho@usp.br

* Correspondence: scozeus@icb.ufmg.br

† These authors contribute to this paper equally.

Received: 27 January 2020; Accepted: 25 April 2020; Published: 28 April 2020

Abstract: The ST2 receptor plays an important role in the gut such as permeability regulation, epithelium regeneration, and promoting intestinal immune modulation. Here, we studied the role of ST2 receptor in a murine model of oral infection with *Brucella abortus*, its influence on gut homeostasis and control of bacterial replication. Balb/c (wild-type, WT) and ST2 deficient mice ($ST2^{-/-}$) were infected by oral gavage and the results were obtained at 3 and 14 days post infection (dpi). Our results suggest that $ST2^{-/-}$ are more resistant to *B. abortus* infection, as a lower bacterial colony-forming unit (CFU) was detected in the livers and spleens of knockout mice, when compared to WT. Additionally, we observed an increase in intestinal permeability in WT-infected mice, compared to $ST2^{-/-}$ animals. Breakage of the intestinal epithelial barrier and bacterial dissemination might be associated with the presence of the ST2 receptor; since, in the knockout mice no change in intestinal permeability was observed after infection. Together with enhanced resistance to infection, $ST2^{-/-}$ produced greater levels of IFN-γ and TNF-α in the small intestine, compared to WT mice. Nevertheless, in the systemic model of infection ST2 plays no role in controlling *Brucella* replication in vivo. Our results suggest that the ST2 receptor is involved in the invasion process of *B. abortus* by the mucosa in the oral infection model.

Keywords: ST2 receptor; *Brucella abortus*; oral infection

1. Introduction

Brucellosis is a worldwide zoonotic disease caused by facultative intracellular pathogen of the genus *Brucella* [1]. Bacteria of the genus *Brucella* infect a wide variety of land and aquatic mammals, including pigs, cattle, goats, sheep, dogs, dolphins, whales, seals, and desert wooden mice. Traditionally, the genus *Brucella* consisted of six recognized species, grouped according to their primary host preferences, i.e.,

B. abortus, bovine; *B. melitensis*, sheep and goats; *B. suis*, pigs; *B. ovis*, sheep; *B. kennels*, dogs; and *B. neotomae*, desert wood mice. Recent new species were isolated from humans (*B. inopinata*), aquatic mammals (*B. pinnipedialis* and *B. ceti*), and from a common rat (*B. microti*), raising the current number to 10 species of the genus [2]. Human brucellosis can be mainly caused by *Brucellaabortus* and *Brucella melitensis*, leading not only to cases of morbidity but also severe economic losses caused mainly by abortions and infertility in infected animals [3]. The natural infection by *Brucella* occurs mainly by the oral and nasal routes through consumption of raw milk and unpasteurized dairy products from infected animals, inhalation of aerosols containing the pathogen, contact with infected animals and their secretions, and by the habit of cattle to lick and smell newborn animals or even aborted fetuses. In addition, there is laboratory and occupational contamination, affecting researchers, farmers, slaughterhouse workers, butchers, and veterinary doctors (many cases of accidental self-inoculation of the vaccine against animal brucellosis), and there are forms (although very unlikely) of human transmission such as contamination of plants by feces and urine from infected animals, and breastfeeding [4–7].

Brucellosis is a systemic disease in which any organ or tissue of the body might be involved. Affected individuals present nonspecific symptoms shared with several other diseases, which cause human brucellosis to present underestimated data of epidemiological distribution [8]. In humans, the main symptoms of the acute phase of the disease are undulating fever, headaches, fatigue, myalgia, and weight loss. In the chronic phase of the disease endocarditis, arthritis, osteomyelitis, and neurological complications can be observed [9]. In animals, brucellosis is a chronic infection that persists throughout life. In females, *Brucella* causes tropism through the bovine placental hormone, erythritol, leading to lesions in the uterine glands, while in males, the bacterium causes tropism through male hormones like testosterone, addressing the testicles. Thus, *Brucella* infection primarily affects the reproductive organs causing abortion and infertility [10].

Brucella spp. can resist death by neutrophils, and replicate within macrophages and dendritic cells, thus, maintaining a long lasting interaction with the host cells [11]. Therefore, innate immunity has developed important mechanisms for detecting and eliminating these bacteria. Toll-like receptors (TLRs) have already proven to be important in the control of *Brucella abortus* infection. The recognition of *B. abortus* molecules by TLR2 (external membrane proteins, Omp16 and Omp19), TLR4 (*Brucella* LPS and *Brucella* lumazine synthase), and TLR9 (*Brucella* DNA), activates intracellular signaling via MyD88, resulting in the activation of NF-κB, MAP kinases, and expression of pro-inflammatory cytokines [12–16]. TLR2 does not participate in the in vivo control of infection, contributing only to the production of pro-inflammatory cytokines [12,14]. However, TLR9 has played a prominent role in relation to in vivo and in vitro control of *B. abortus* infection [17]. In addition to the receptors mentioned above, *B. abortus* leads to activation of NLRP3 (through reactive mitochondrial oxygen species induced by bacteria) and AIM2 (recognition of bacterial DNA) inflammasomes, leading to activation of innate immunity and infection control [18]. The STING protein was also determined as an important adapter molecule required for resistance against this bacterium [19,20].

In the context of intestinal immunity, it is important to highlight the importance of these receptors in this microenvironment, since several of these innate immunity receptors such as the TLRs, NRLs, G protein-coupled receptors (GPCRs), and STING are also expressed in the intestinal mucosa, having an important function in the maintenance of host commensal microbiota and intestinal homeostasis [21–23]. Among these innate immunity receptors, the ST2 receptor and its ligand, cytokine IL-33 has been widely studied since its discovery [24]. The ST2 receptor of the IL-1 family, also called IL1rl1, tumorigenicity suppressor 2, growth stimulation expressed in gene 2 and serum stimulation 2, was classified as a receptor for IL-33 in 2005 [25,26].There are four isoforms encoded by the ST2 gene. The two most prominent isoforms include the ST2L transmembrane, which acts as a membrane receptor, responsible for binding the IL-33 and activating the signaling cascades to improve the functions of the cells that express this receptor, and the sST2, presented in a soluble form, which acts by sequestering the free IL-33, preventing its signaling. They are the consequence of a double system of promoters (sST2 proximal promoter and ST2L distal promoter) that results in the differential expression of mRNA.

ST2L, like other IL-1 receptors, consists of an extracellular domain, transmembrane domain and cytoplasmic domain (Toll/interleukin-1 receptor (TIR)), while sST2 does not have the transmembrane and cytoplasmic domains and therefore exists as a soluble protein. In addition, alternative splicing results in the formation of ST2V and ST2LV. ST2V shares the same extracellular and transmembrane domain as ST2L, but is remarkable for its unique hydrophobic tail and is particularly enriched in the gastrointestinal tract. Finally, ST2LV notably does not have the ST2L transmembrane domain, but maintains the intracellular domain [27,28].The ST2 receptor is expressed in a wide variety of immune cells, such as conventional T cells, particularly regulatory T cells (T regs) [29], innate type 2 lymphoid cells (ILC2) [30], polarized macrophages M2 [31], eosinophils [32], basophils [33], neutrophils [33], NK cells [34], iNKT cells [34], and several other immune cells and their soluble isoform.sST2 can be produced spontaneously by the small intestine.

As an alarmine, IL-33 is one of the first molecules that "sounds the alarm" to indicate that there has been a violation of the primary defenses of the intestinal epithelium against pathogens and other threats [35]. IL-33 is produced by a variety of stromal cells and organ parenchyma, such as smooth muscle cells, fibroblasts, myofibroblasts, endothelial cells, glia cells, osteoblasts, adipocytes, and by different cells of the immune system, such as macrophages, dendritic cells, and mast cells [36]. IL-33 acts on several cell types, including cells of hematopoietic origin and non-hematopoietic cells. The secretion of IL-33 has been described in monocyte lineage (THP-1 cells), in response to different stimuli—bacterial infection, lipopolysaccharide (LPS) with aluminum adjuvant, and isolated LPS. [29,32]. In order to maintain the integrity of this mucosal barrier, the intestinal epithelium undergoes rapid and continuous self-renewal to replace the damaged cells. Activation of the IL-33/ST2 pathway in epithelial progenitor cells leads to inhibition of Notch signaling and results in differentiation of stem cells towards a line of secretory intestinal cells [37], resulting in the production of mucin, an important barrier mechanism of intestinal immunity, decreasing the interaction of the intestinal epithelium and pathogenic bacteria [38]. Moreover, the activation of this axis is important to recruit and activate innate immune cells, inducing Th1 or Th2 responses, according to the required immune response [39]. Although brucellosis is a worldwide zoonosis, the mechanisms involved during the course and establishment of the natural oral infection by *Brucella abortus* are still poorly studied. With regard to the process of invasion of *Brucella* through mucosal barriers, there are few studies on the mechanisms involved in the ability of this pathogen to interact with the epithelial cells of the gastrointestinal (GI) tract with the host microbiota, and also with the subsequent immune and homeostatic response in the gastrointestinal tract. The intraperitoneal infection pathway is the most commonly used in studies using the murine model. This route favors the immediate systemic dissemination of *Brucella* and its proliferation in lymphoid tissues, especially in the spleen. However, considering that the oral route is the main route of natural infection in humans and animals, there is a need to understand the mechanisms of the establishment of oral infection, so new therapeutic strategies can be developed in order to control this disease. Since, the IL-33/ST2 axis is positioned to interact with the main components of the intestine, which include epithelial cells in response to cell damage and a microbiome composed of commensal bacteria and immune mucosal cells [24], we investigated the role of the ST2 receptor in the immune response against *Brucella abortus* oral infection.

2. Results

2.1. The Absence of the ST2 Receptor Confers Partial Resistance to Oral Infection

Considering that one of the main routes of the *Brucella* infection is through oral surfaces, we assessed the susceptibility of wild-type (WT) mice and animals deficient for the ST2 receptor ($ST2^{-/-}$) to oral infection with *Brucella abortus*, by determining the number of colony forming units in livers and spleens, 3to 14 days post-infection. We observed higher CFUs of *Brucella* in livers (Figure 1A) and spleens (Figure 1B) of WT mice compared to $ST2^{-/-}$ animals, 3 days after oral infection. Regarding the time of 14 days post-infection, we also observed reduced numbers of bacterial CFUs in livers of $ST2^{-/-}$

mice compared to WT, but not in the spleens of these animals. These findings suggest an enhanced resistance to *Brucella* infection in ST2 knockout mice compared to WT.

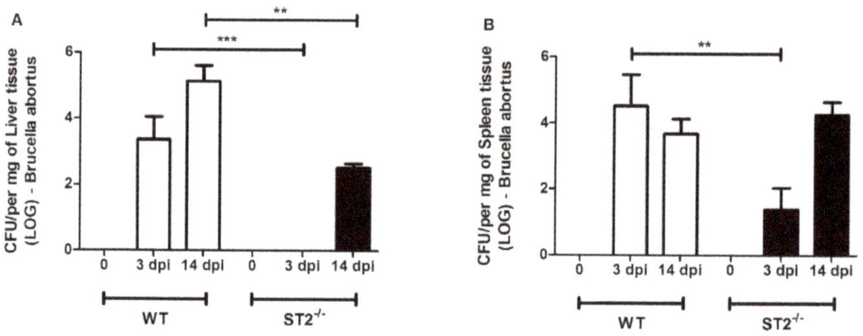

Figure 1. The absence of the ST2 receptor confers partial resistance to oral infection. Wild-type (WT) mice and ST2-deficient mice were orally infected by 1×10^9 colony-forming unit (CFU) of *Brucella abortus* and were sacrificed after 3 and 14 days of infection. The livers (**A**) and spleens (**B**) of the mice were collected and processed for evaluation of the number of viable bacteria through CFU counts. Results expressed as mean ± standard deviation (n = 5–7). The data are representative of 3 independent experiments. ** $p < 0.01$, *** $p < 0.001$.

2.2. *Absence of ST2 Resulted in Change of Intestinal Architecture*

Intestinal epithelial cells produce antimicrobial effectors that play a central role in shaping the gut microbial community and protecting mucosal tissues from colonization and invasion of commensal microorganisms. To investigate the potential role of ST2 in intestine homeostasis, we analyzed histology sections of small intestine from WT and $ST2^{-/-}$ mice. First, we observed in H&E-stained sections that the villi, crypt, and mucosa thickness of small intestine in naive $ST2^{-/-}$ mice were shorter than in WT animals (Figure 2A–C), regardless of the infection. After oral infection by *Brucella abortus*, we did not observe a major alteration in the gut architecture within each mouse group. Representative photomicrographies of hematoxylin-and-eosin-stained duodenum sections from WT (Figure 2D) and $ST2^{-/-}$ (Figure 2E) are shown. Together, these data indicate that an intact ST2 signaling is important to maintain gut mucosa integrity.

2.3. *ST2 Receptor is Important in the Maintenance of the Intestinal Epithelial Barrier*

The role of the ST2 receptor in maintaining the integrity of the intestinal epithelial barrier following *Brucella infection* was evaluated by the FITC-labeled dextran flow method (Figure 3A). We observed that in the WT animals, *Brucella* infection led to an increased permeability of the epithelial barrier (Figure 3A) (observed by increase of FITC-dextran in the serum of animals). In contrast, $ST2^{-/-}$ mice intestinal permeability was not altered after infection (3 days). We also evaluated the regulation of amphiregulin (AREG) and mucin molecule 2 (MUC2) expression in WT and $ST2^{-/-}$ mice, after 3 days of oral infection with *Brucella*. Amphiregulin and MUC2 are two important components to protect the intestinal epithelium. Regarding the expression of amphiregulin, which is critical for intestinal epithelial regeneration after injury, *B. abortus* infection increased the expression of *AREG* in both WT and $ST2^{-/-}$ mice (Figure 3B). Additionally, we determined that *MUC2* expression following *B. abortus* infection requires ST2 (Figure 3C), suggesting the participation of ST2 in the transcriptional regulation of this molecule. Tight junctions (TJs) play an important role in intestinal function. TJs in intestinal epithelial cells are composed of different junctional molecules, such as claudins, zonula occludens (ZO-1, -2, and -3), among others. Therefore, we determined the role of ST2 in *ZO-1*, *-2*, and *-3* and *claudin-1* expression in intestinal tissue. Herein, we showed that animals lacking ST2 had reduced expression levels of *ZO-1* and to a less extent, that of *ZO-2* and *-3*, when compared to WT mice

(Figure 3D,E,F). Regarding *claudin-1* mRNA transcripts, the levels of this TJ remained similar between both mouse groups (Figure 3G).

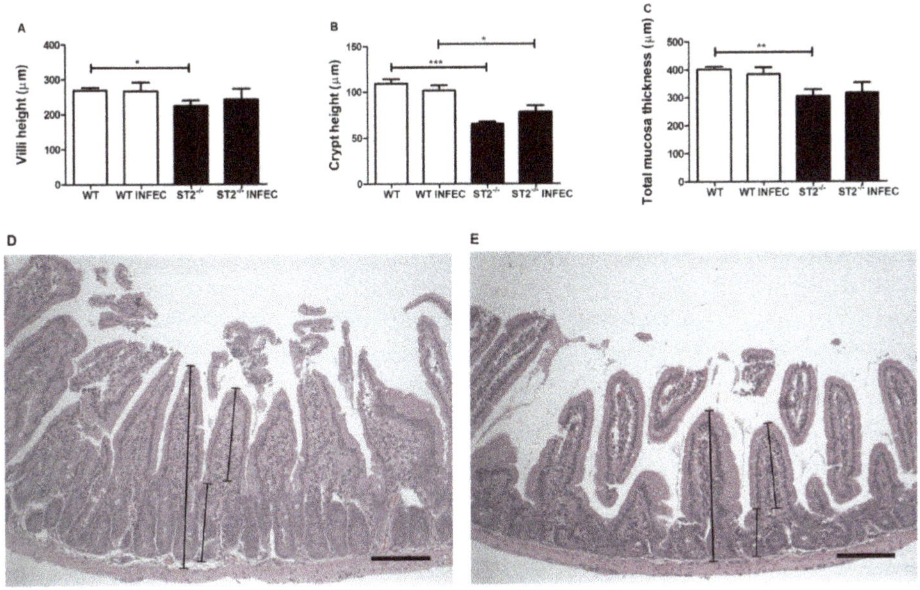

Figure 2. Alterations in mucosa structure in WT and ST2$^{-/-}$ mice during *Brucella* infection. Duodenum of wild-type (WT) and ST2$^{-/-}$ uninfected and infected mice were collected for analysis of (**A**) villi height, (**B**) crypt height and (**C**) total mucosa thickness. Representative photomicrographies of hematoxylin and eosin–stained duodenum sections from WT (**D**) and ST2$^{-/-}$ (**E**) mice evidencing total mucosa thickness, crypt and villi height. Bars represent 100 μm. * $p < 0.05$; ** $p < 0.01$; *** $p < 0.001$.

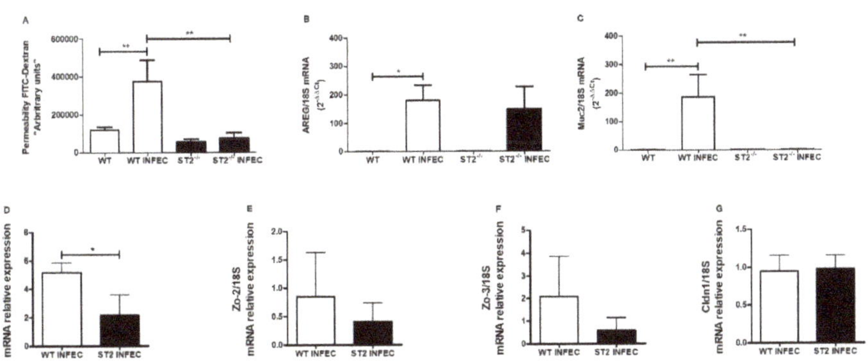

Figure 3. The ST2 receptor is important in the maintenance of the intestinal epithelial barrier and in the transcriptional regulation of Muc2 and ZO-1. WT mice and ST2-deficient mice were orally infected with 1×10^9 CFU of *B. abortus* and after 3 days of infection intestinal permeability was evaluated (**A**). Small bowel samples were also collected for transcriptional analysis of *AREG* (**B**), *Muc2* (**C**), *ZO-1* (**D**), *ZO-2* (**E**), *ZO-3* (**F**), and *claudin-1* (**G**) genes, using qPCR. Results expressed as mean ± standard deviation (n = 5–7). The data are representative of two experiments. * $p < 0.05$; ** $p < 0.01$.

2.4. Lack of ST2 Receptor Modulates the Recruitment of Neutrophils and Eosinophils and Increases the Production of IFN-γ and TNF-α in Small Intestine after *Brucella abortus* Infection

In order to investigate whether the inflammatory response could be involved in the resistance phenotype observed in ST2$^{-/-}$ after *B. abortus* infection, we determined myeloperoxidase (MPO) and eosinophilic peroxidase (EPO) activity as an indirect measurement of neutrophils and eosinophils influx. After *Brucella* infection, there was an increase in MPO (Figure 4A) and EPO (Figure 4B) activity in WT mice, which was not observed in the ST2$^{-/-}$-infected animals, suggesting that the absence of the ST2 receptor somehow modulated the recruitment of neutrophils and eosinophils after infection. Additionally, we also determined the participation of ST2 in the production of cytokines involved in the intestinal immune response, such as IFN-γ, TNF-α, IL-10, IL-1β, and IL-33. The level of these cytokines was measured in small bowel fragments, from non-infected and infected mice after 3 days of infection. Herein, we observed that *Brucella* infection increased the production of IFN-γ (Figure 4C) and TNF-α (Figure 4D) in ST2$^{-/-}$ mice, when compared to the WT animals. This Th1-like profile detected in ST2$^{-/-}$ mice might be related to a reduction in the bacterial load observed in the spleens and livers of these animals, as previously observed by us and others [40,41]. Regarding the production of IL-10 (Figure 4E) and IL-1β (Figure 4F), there is no difference in the levels of these cytokines produced between ST2$^{-/-}$-infected mice when compared to the WT-infected animals. As for IL-33 (Figure 4G), an ST2-binding cytokine, we observed that in ST2$^{-/-}$ mice the production of this cytokine was already naturally decreased and after infection there was no change in this profile when compared to the WT mice.

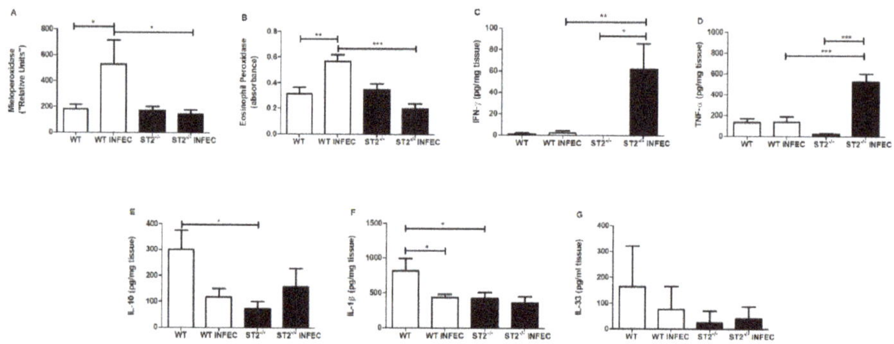

Figure 4. ST2 receptor deficiency modulates the recruitment of neutrophils and eosinophils and increases the production of IFN-γ and TNF-α after *Brucella abortus* infection. WT and ST2$^{-/-}$ mice were infected orally with 1×10^9 CFU of *B. abortus* and after 3 days of infection, small intestine samples were collected for processing and evaluation of myeloperoxidase (**A**) and eosinophil peroxidase (**B**). Tissue samples were also assessed for cytokine production, such as IFN-γ (**C**), TNF-α (**D**), IL-10 (**E**), IL-1β (**F**), and IL-33 (**G**) by ELISA. Results are expressed as mean $\pm$ standard deviation (n = 5–7). The data are representative of 3 independent experiments. * $p < 0.05$; ** $p < 0.01$; *** $p < 0.001$.

2.5. ST2 Receptor Does Not Play a Role in Systemic Infection Caused by *Brucella abortus*

The resistance or susceptibility phenotype to systemic infection by *Brucella abortus* was evaluated by determining the number of CFU in the livers and spleens of WT versus ST2$^{-/-}$mice, after 3 and 14 days of intraperitoneal (i.p.) infection. We observed that the bacterial load was similar in livers (Figure 5A) and spleens (Figure 5B) of WT and ST2$^{-/-}$ mice after infection. These findings suggest that lack of ST2 plays no role in *Brucella* control in vivo, after intraperitoneal infection.

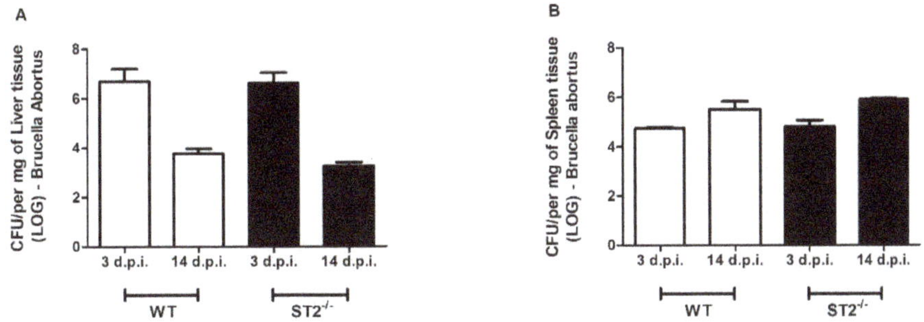

Figure 5. Lack of ST2 receptor does not influence systemic infection induced by *Brucella abortus*. WT mice and ST2-deficient mice were infected intraperitoneally with 1×10^6 CFU of *B. abortus* and sacrificed after 3 and 14 days of infection. The livers (**A**) and spleens (**B**) of these mice were collected and processed for evaluation of the number of viable bacteria through CFU count. Results expressed as mean ± standard deviation (n = 5–7). The data are representative of 3 independent experiments.

2.6. The Absence of the ST2 Receptor Does Not alter the Production of Nitric Oxide by Macrophages

To evaluate the nitric oxide (NO) production in WT and $ST2^{-/-}$ macrophages, and to correlate it with the potential microbicide activity, nitrite, a stable metabolite of NO, was measured using Griess reagent on macrophage supernatants. We observed that the production of NO in the macrophages of both mouse strains when stimulated with *Brucella* or LPS is similar, in the presence or absence of IFN-γ (Figure 6). Therefore, our findings suggest that ST2 deficiency does not influence the ability of *Brucella* infected macrophages to produce NO.

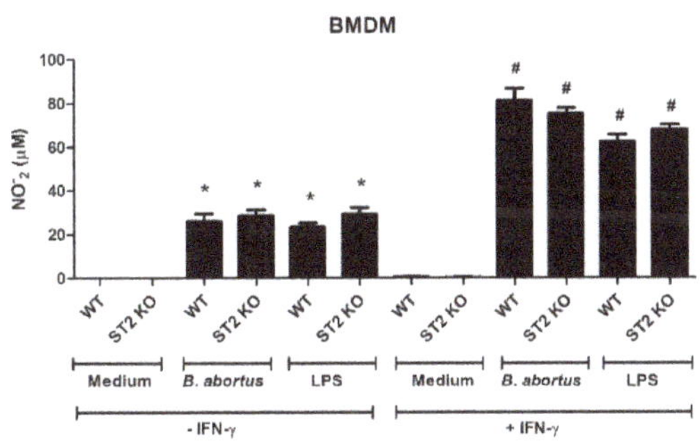

Figure 6. The absence of the ST2 receptor did not alter the production of nitric oxide by macrophages. Macrophage was derived from WT and ST2-deficient mice bone marrow and stimulation with *Brucella abortus* or lipopolysaccharide (LPS) was performed in the presence or absence of IFN-γ. The supernatant was collected to perform the Griess assay, as already described. * $p < 0.001$ when compared to the medium. # $p < 0.001$ when compared to the cells with no IFN-γ. Results expressed as mean ± standard deviation (n = 5–7). The data are representative of two independent experiments.

3. Discussion

Infections caused by the bacteria of the genus *Brucella* were mainly transmitted orally to human and animals. *Brucella* has a rapid capacity for infectivity in the oral infection model (by gavage or inoculation in the oral cavity), where after one hour of infection, bacteria were already found in the lumen and in the epithelium of the duodenum [42]. Few virulence factors of *Brucella* that are important for the establishment of infection through the oral route have been described, such as urease [43], which confers resistance to gastric acidity, cholylglycine hydrolase (CGH) [44] which induces resistance to bile salts, and the *Brucella* protease inhibitor Omp19 [42] which induces resistance to the action of proteases.

When gastrointestinal tract defense cells fail to capture microorganisms, they are drained mainly through the portal vein into the liver [45]. Previous studies have shown that in bacterial infections, higher concentrations of LPS are detected in the portal vein when compared to other hepatic or peripheral veins and, interestingly, bacteria can be cultivated even from healthy liver explants [46,47]. The phenotype of resistance exhibited by ST2-receptor deficient mice during oral infection was lost when intraperitoneal infection was performed. Considering that the portal vein might be the main route for systemic dissemination of *Brucella*, we first speculated that the liver from ST2-deficient mice might be mounting its own immune response and consequently, decreasing the number of viable bacteria and their ability to spread systemically. However, when we measured IFN-γ and TNF-α production by liver cells, ST2 knockout and WT mice produced similar levels of these cytokines (Figure S1 found in the Supplementary Materials). Therefore, we suggest that other mechanisms might be involved in reduced bacterial counts observed in ST2$^{-/-}$ livers.

The gut-associated lymphoid tissue (GALT), such as the Peyer's patches (PPs) along with the intestinal mucosal epithelium, act as a sentinel for recognition and initiation of immune responses against pathogenic bacteria [48]. The process of invasion of *Brucella* into the gastrointestinal tract occurs through its ability to translocate via M cells, which occurs by interaction with the prionic protein PrP^{c} that are highly expressed on the apical surface of these cells [49]; however, this process does not lead to the rupture of the cell–cell junctions [50]. Another mechanism related to the invasion process is through the intestinal epithelial cells [51], but this mechanism has not yet been fully clarified. Tight junctions (TJs) play an important role in intestinal function. TJs in intestinal epithelial cells are composed of different junctional molecules, such as claudins, zonula occludens (ZO-1, -2, and -3), and occluding, among others. In this study, we determined the role of ST2 in *ZO-1, -2,* and *-3* and *claudin-1* expression in intestinal tissue. Herein, we showed that animals lacking ST2 had reduced expression levels of *ZO-1* and to a less extent that of *ZO-2* and *ZO-3*, when compared to WT mice. Regarding *claudin-1* mRNA transcripts, the levels of this TJ remained similar between both mouse groups. The reduced expression of zonula occludens (ZO) molecules might not have a direct relationship to intestinal permeability in this model since ST2$^{-/-}$ mice had reduced intestinal permeability, compared to WT animals, as measured by the FITC-dextran method. Rather, diminished expression of these tight junction gene products might correlate with enhanced IFN-γ production observed in ST2$^{-/-}$ animals, as recently demonstrated in the *Salmonella enteritis* infection model [52]. Breaking the epithelial barrier after oral infection resulted in increased intestinal permeability observed in WT mice and could be one important mechanism that facilitates the entry and spread of this pathogen. Studies using a model of ex vivo infection in the ileal bowel loop showed that the migration of *Brucella* through the intestinal epithelium occurs via endocytosis by the follicle-associated epithelium (FAE) in Peyer's patches or by its uptake by the penetrating dendritic cells of the FAE [49,53]. Additionally, Rosseti and collaborators (2013) [54] observed through microarray analysis that two pathways related to the intestinal epithelial barrier were repressed during the initial phase of *Brucella* infection, suggesting the subversion of the barrier function and facilitating transepithelial migration. Thus, a variety of pathogens use molecules involved in cell adhesion and invasion, such as the *Helicobacter pylori*, whose type IV secretion system injects one of its effectors (CagA) into the host cell, modifying several processes and culminating in the rupture of the epithelial barrier and invasion of the bacteria [55]. Although, we did not explore the infection of *Brucella abortus* in intestinal epithelial cells in vitro, the breaking of the epithelial barrier in vivo might be associated with the presence of the ST2 receptor, since no increase in intestinal permeability was observed in ST2 knockout mice after infection, when compared to WT animals.

Another mechanism related to the process of maintaining the epithelial barrier, involves amphiregulin (AREG), which plays a role in intestinal epithelial regeneration after injury [56] and in cellular proliferation [57]. The level of expression of this molecule after infection was similar in both animals analyzed, suggesting that the change in intestinal permeability observed in WT mice is not mediated by the participation of ST2 in transcriptional regulation of *AREG*. The intestinal mucus is one of the main components of defense against invasion of pathogens and protects the epithelium from physical damage. Muc2 mucin is produced and secreted by intestinal goblet cells. We believe that the increased expression of MUC2 in WT mice might be linked to augmented intestinal permeability. Recently, a higher expression level of the mucin glycoprotein Muc2 in enteroids following *Shigella flexneri* infections was reported [58]. These findings suggest that mucus production might not be an important factor involved in the phenotype of decreased intestinal permeability observed in ST2 knockout-infected mice, and that other mechanisms are involved in the intestinal barrier of $ST2^{-/-}$ animals.

During infections, depending on the organ involved, IL-33/ST2 signaling might induce the necessary immune response to control the infectious foci, which might be a Th1 or Th2 type of response [38]. The increase of IFN-γ and TNF-α after the infection observed in the knockout mice might be associated with a greater defense of the intestine against the invasion of *B. abortus*, which might be contributing to the phenotype of resistance observed in these animals. Previous studies have demonstrated the requirement of Th1-type cytokine profile to induce protection against *Brucella* infection [40,41]. IL-1β plays an important role in the defense against pathogens and also in the maintenance of the intestinal homeostatic balance and in the regeneration of the epithelium [59,60]. High levels of this cytokine are found in the intestinal mucosa in a normal state (steady-state), implying its importance in maintaining the mucosal barrier and in immune monitoring. The decrease in IL-1β levels observed in WT-infected mice corroborates the data of altered intestinal permeability through infection, suggesting that ST2 might have a regulatory role of this cytokine and consequently a function in the maintenance of intestinal permeability. Several studies have proposed that cell injury or death are the dominant mechanisms through which the IL-33 reaches the extracellular environment. Therefore, in a steady state, the IL-33 is not actively secreted by cells [35,61], being an important tool for the immune system, when there is a violation in the integrity of the mucosa, secondary to damage to the epithelial cells [61]. In this study, we observed that production of this cytokine in the small intestine was naturally higher in WT mice, compared to the knockout animals, suggesting that the change in intestinal permeability induced by oral infection was not through tissue damage, but via other mechanisms that need to be investigated.

The increase in MPO and EPO as an indirect measurement of neutrophils and eosinophils in the intestine of WT mice might contribute to the establishment of infection. Since *Brucella abortus* is an intracellular pathogen, and it is already described in the literature that neutrophils infected with *Brucella* are readily phagocytized by macrophages and replicate extensively within these cells, neutrophils then end up serving as "Trojan horse" vehicles for efficient bacterial dispersion, intracellular replication, and establishment of chronic infections [62]. In ST2 knockout mice, MPO and EPO were decreased after infection when compared to WT animals, which might contribute to the initial resistance profile exhibited by these mice, since there are less infected granulocytes that can carry the pathogen to spread into other host cells and organs. The absence of ST2 increased the bactericidal activity of neutrophils and macrophages against *Staphylococcus aureus* in a sepsis model [63], by increasing the production of nitric oxide of these cells. Thus, we sought to investigate whether, in an in vitro scenario, macrophages would show higher production of NO against *Brucella* infection. We observed that nitric oxide production rate is similarly influenced in WT and $ST2^{-/-}$ macrophages, either through stimulation with *B. abortus* or LPS, in the presence or absence of IFN-γ, suggesting that bone marrow-derived macrophages have the same microbicidal potential, and that ST2 in the context of *B. abortus* infection is not involved in the regulation of NO production by these cells.

In summary, we observed that lack of ST2 is important in the model of *Brucella* oral infection but not when the animals are infected by the intraperitoneal route. In this study, we revealed that the oral infection by *Brucella abortus* alters the intestinal homeostasis in favor of its invasion and establishment of systemic infection, and the mechanisms involved in this process were partially dependent on the ST2 receptor. The ST2 receptor proved to be important in maintaining the epithelial barrier and in the negative regulation of the inflammatory immune response to oral infection through *B. abortus*.

4. Materials and Methods

4.1. Mice

Wild-type Balb/C (WT) mice were purchased from the Federal University of Minas Gerais (UFMG), and ST2 KO (kindly provided by Dr. José Carlos Alves-Filho, Department of Pharmacology, Ribeirao Preto Medical School, University of Sao Paulo, Brazil). Genetically deficient and control mice were maintained at our facilities and used at 6–8 weeks of age. Mice were housed in filter-top cages and provided with sterile water and food ad libitum. Groups of 5 to 7 animals were used to perform all experiments. The procedures for animal experimentation were approved by the Ethics Committee for the Use of Animals of the Federal University of Minas Gerais—CEUA/UFMG under protocol number 273/2017.

4.2. Bacteria

Brucella abortus smooth virulent strain 2308 was obtained from our laboratory collection. Frozen stocks were prepared from isolated colonies previously grown in *Brucella* broth medium (BB) + 1.5% agar for 3 days. One day prior to infection, *B. abortus* was grown in liquid BB and the OD was measured in a spectrophotometer. In all experiments performed in this study, OD_{600} 1 = 3 × 10^9 CFU/mL.

4.3. Bacterial Counting in B. abortus Infected Mice

Five to seven mice from each group (Balb/c or $ST2^{-/-}$) were infected orally by intragastric gavage with 1 × 10^9 or intraperitoneally (i.p.) with 1 × 10^6 virulent *B. abortus* S2308 in 100 µL of PBS. After 3 or 14 days post-infection, mice were sacrificed and liver and spleens were used to determine the number of bacteria through CFU counting. All organs harvested from each animal were weighed and macerated in saline (NaCl 0.9%). To determine bacterial burden, livers and spleens were serially diluted in saline and plated in duplicates on BB agar. Plates were incubated for 3 days at 37 °C and the CFU number was determined.

4.4. Intestinal Permeability Assay

The in vivo intestinal permeability assay to verify the barrier function was performed using the FITC-labeled Dextran method with minor modifications [64]. Briefly, food and water were removed and, after 3 h, mice were weighed and received intragastric inoculation of FITC-Dextran (0.6 mg/g body weight, PM 4000; Sigma-Aldrich, St. Louis, MO, USA). Four hours after gavage, the animals were anesthetized with ketamine/xylazine (Syntec, São Paulo, Brazil) (0.6 mL of ketamine at the concentration of 100 mg/mL, 0.4 mL of xylazine at the concentration of 20 mg/mL, and 4 mL of saline), blood was taken by cardiac puncture and was subsequently euthanized. Blood was centrifugedat 10,000 rpm for 3 min at 4 °C and serum collected was pipetted in the volume of 100 µL/well, in a plate of 96 wells (Nunc, Thermo Fisher Scientific, Norcross, GA, USA). The measurement of the fluorescence intensity of each sample (excitation, 492 nm; emission 525 nm; Synergy2, Bio Tek Instruments, Inc., Winooski, VT, USA) was performed. The measurement of intestinal permeability was expressed as the mean of the fluorescence unit. Increased fluorescence in the serum indicated increased intestinal permeability.

4.5. Measurement of Myeloperoxidase (MPO) and Eosinophilic Peroxidase Activity (EPO)Activity

The evaluation of the MPO and EPO enzyme activity was used as an indirect index of neutrophil and eosinophil recruitment in the tissues, respectively. The protocol for dosage of this enzyme in homogenized tissues was performed with some modifications [65]. In brief, fragments of small intestine (100 mg) of the animals were removed and frozen at –80 °C. After thawing, the tissue was homogenized in 4.7 pH buffer (0.1 M NaCl, 0.02 M $NaH_2PO_4.1H_2O$, 0.015 M Na_2-EDTA) (100 mg of tissue in 1.0 mL buffer), using a tissue homogenizer, centrifuged at 10,000 rpm for 15 min at 4 °C and the precipitate was submitted to hypotonic lysis (500 µL of 0.2% NaCl solution followed by addition of equal volume of solution containing 1.6% NaCl and 5% glucose, 30 s after) for RBC lysis. After further centrifugation, the precipitate was resuspended in 0.05 M NaH_2PO_4 buffer (pH 5.4) containing 0.5% hexadeciltrimethylammonium bromide (HTAB) (Sigma) and was re-homogenized. Aliquots of 1 mL of the suspension were transferred to microcentrifuge tubes of 1.5 mL and submitted to three freezing/thawing cycles using liquid nitrogen. These samples were again centrifuged for 15 min at 10,000 rpm. The supernatant was collected and MPO activity was calculated by measuring the changes in optical density (OD) at 450 nm, using tetramethylbenzidine (TMB) (1.6 mM) (Sigma) and H_2O_2 (0.5 mM). The supernatant was also used to quantify the peroxidase activity. The assay was performed in 96-well plates, 75 µL per sample or blank well (PBS/HTAB 0.5%) was incubated with 75 µL of substrate (o-phenylenediamine (OPD) (Sigma) 1.5 mM, in Tris-HCl buffer—0.075 µM, pH 8, supplemented with H_2O_2 6.6 mM). The plate was incubated at 20 °C in the dark for approximately 30 min and the reaction was interrupted by the addition of 50 µL of H_2SO_4 1M. The reaction was measuredin a microplate reader (Multiskan FC Thermo Scientific, Norcross, GA, USA) with a 492 nm filter.

4.6. Measurement of Cytokine Concentrations

To evaluate the production of cytokines, fragments of the small intestine with approximately 100 mg were homogenized using a tissue homogenizer (T10 Basic ULTRA-TURRAX®, IKA, Königswinter, Germany) in 1 mL of cytokine extraction solution—PBS containing antiprotease cocktail (0.1 mM PMSF, 0.1 mM benzethonium chloride, 10 mM EDTA, and 20 KI aprotinin A) and 0.05% Tween-20. Then, the homogenates were centrifuged at 4 °C for 10 min at 10,000 rpm. The supernatants were immediately collected and stored at −80 °C for subsequent measurement. The concentrations of IL-1β, IL-33, IFN-γ, TNF-α, and IL-10 was performed through the ELISA method, using kits purchased from R&D Systems (DuoSet) (R&D Systems, Minneapolis, MN, USA) according to manufacturers' recommendations.

4.7. Real-Time PCR (RT–PCR)

RNA was extracted from small intestine with TRIzol reagent (Invitrogen, Thermo Fisher Scientific, Norcross, GA, USA) according to the manufacturer's instructions. cDNA was synthesized by reverse transcription (RT) from 1 µg of total RNA and was used to perform RT–PCR in a final volume of 10 µL containing SYBR green PCR Master Mix (Applied Biosystems, Carlsbad, CA, USA) and 20 µM of primers. RT–PCR was performed in triplicates, on an ABI 7900 Real-time PCR system (Applied Biosystems). The primers used for gene amplification were as follows: 18S forward 5′-CGTTCCACCAACTAAGAACG-3′, reverse 5′- CTCAACACGGGAAACCTCAC-3′; MUC2 forward 5′- CACCAACACGTCAAAAATCG -3′,reverse 5′- CGCAGAACTCCCAGTAGCA -3′; Amphiregulin forward 5′- GCCATTATGCAGCTGCTTTGGAGC -3′, reverse 5′- TGTTTTTCTTGGGCTTAATCACCT -3′; ZO-1 forward 5′-TGAACGCTCTCATAAGCTTCGTAA-3′, reverse 5′-ACCGTACCAACCATCATTC ATTG-3′; ZO-2 forward 5′-CCATGGGCGCGGACTATCTGA-3′, reverse 5′-CTGTGGCGGGGAGGTT TGACTTG-3′, ZO-3 forward 5′-AAGCACGCAATCCTGGATGTCACC-3′, reverse 5′-GTCGCG CCTGCTGTTGCTGTATTA-3′; claudin-1 forward 5′-AGCCAGGAGCCTCCCCGCAGCTGCA-3′, reverse 5′-CGGGTTGCCTGCAAAGT-3′. The levels of mRNAs are presented as relative expression units after normalization to 18S transcripts.

4.8. Generation of BMDMs

Bone-marrow cells were obtained from femur and tibiae of ST2 KO and WT mice and they were differentiated into BMDMs using a previously described protocol, with some modifications [66]. In brief, cells were seeded on 24-well plates at 5×10^5 cell/mL (day 0) and maintained in DMEM medium containing 10% FBS, 100 U/mL penicillin, 100 µg/mL streptomycin, and 20% LCCM (L929-conditioned medium), at 37 °C in a 5% CO_2 atmosphere for 7 days. On day 4 of incubation, the medium was fully replaced. Four hours before stimulation or infection, BMDMs were maintained only in the DMEM medium containing 1% FBS.

4.9. Nitrite Measurement by Griess Reagent

The nitric oxide assay was performed as described previously [15]. The concentration of nitrite (NO_2^-), a stable metabolite of NO, was measured using Griess reagent (1% sulfanilamide and 0.1% naphthylethylenediaminedihydrochloride in 2.5% phosphoric acid). In brief, 50 μL of cell culture supernatants was mixed with 50 μL of Griess reagent. Subsequently, the mixture was incubated, protected from light at room temperature for 5 min, and the absorbance at 550 nm was measured in a microplate reader. Fresh culture medium (DMEM + 1% FBS) was used as a blank in every experiment. The quantity of nitrite was determined from a sodium nitrite ($NaNO_2$) standard curve.

4.10. Gut Pathology

The small intestine of the animals was removed soon after the sacrifice, and the duodenum was separated for histological analysis. The tissues were extended in contact with the filter paper and opened by removing all their contents without damaging the mucosa. The fragments were transferred to a container containing 10% formaldehyde solution for a short period for pre-fixing. The prefixed material was placed on a flat surface and wound in a spiral with the mucosa facing inwards to form rolls. The rolls were tied with line and fixed by immersion in 10% formalin solution in PBS, pH 7.4 for 48 h, and embedded in paraffin. One 4-μm-thick sections were obtained and stained with hematoxylin-and-eosin (H&E) and examined under light microscopy by two pathologists blinded to the experiment. Measurement of villus heights, crypt, and total mucosa thickness depth was performed using the ImageJ software. Fifteen intact and well-oriented villi, crypts, and total mucosa thickness were measured from each animal of each mouse group ($n = 5$).

4.11. Statistical Analysis

The experiments were repeated at least twice with similar results. Graphs and data analysis were performed using GraphPad Prism 5 (GraphPad Software, San Diego, CA, USA), using one-way ANOVA followed by a post-test of Student-Newman-Keuls.

Supplementary Materials: The following are available online at http://www.mdpi.com/2076-0817/9/5/328/s1, Figure S1: ST2 receptor deficiency does not influence the production of IFN-γ and TNF-α in liver after *Brucella abortus* infection.

Author Contributions: Conceptualization, R.S., A.T.V. and S.C.O.; Data curation, P.C.C., M.R., D.S. and F.V.M.; Formal analysis, P.C.C., J.C.A.-F., A.T.V. and S.C.O.; Funding acquisition, S.C.O.; Investigation, R.S., P.C.C., M.R., V.R., D.S., V.M., F.V.M., F.M., M.F.R., D.C.d.R., G.D.C. and J.C.A.-F.; Methodology, R.S., P.C.C., M.R., V.R., D.S., V.M., F.V.M., M.F.R., J.C.A.-F., and A.T.V.; Project administration, A.T.V. and S.C.O.; Resources, S.C.O., F.M., M.F.R. and J.C.A.-F.; Supervision, A.T.V. and S.C.O.; Validation, J.C.A.-F. and A.T.V.; Writing—original draft, R.S., A.T.V. and S.C.O. All authors have read and agreed to the published version of the manuscript.

Funding: This study was carried out with the financial support of CNPq (grants# 302660/2015-1 and 406883/2018-1), FAPEMIG (grants# APQ-00837-15 and APQ-01945-17) and National Institute of HealthR01 AI116453.

Conflicts of Interest: The authors declare no conflict of interest.

References

1. Moreno, E.; Moriyon, I. *Brucella melitensis*: A nasty bug with hidden credentials for virulence. *Proc. Natl. Acad. Sci. USA* **2002**, *99*, 1–3. [CrossRef] [PubMed]
2. De Figueiredo, P.; Ficht, T.A.; Rice-Ficht, A.; Rossetti, C.A.; Adams, L.G. Pathogenesis and immunobiology of brucellosis: Review of Brucella-host interactions. *Am. J. Pathol.* **2015**, *185*, 1505–1517. [CrossRef] [PubMed]
3. Moreno, E. Retrospective and prospective perspectives on zoonotic brucellosis. *Front. Microbiol.* **2014**, *5*, 213. [CrossRef] [PubMed]
4. Yoo, J.R.; Heo, S.T.; Lee, K.H.; Kim, Y.R.; Yoo, S.J. Foodborne outbreak of human brucellosis caused by ingested raw materials of fetal calf on Jeju Island. *Am. J. Trop. Med. Hyg.* **2015**, *92*, 267–269. [CrossRef]
5. Seleem, M.N.; Boyle, S.M.; Sriranganathan, N. Brucellosis: A re-emerging zoonosis. *Vet. Microbiol.* **2010**, *140*, 392–398. [CrossRef]

6. Silva, F.L.; Paixao, T.A.; Borges, A.M. Brucelose Bovina. *Cad. Tec. Vet. Zoot.* **2005**, *47*, 1–12.
7. Secretaria de Estado da Saúde do Paraná (Ed.) *Protocolo de Manejo Clínico e Vigilância em Saúde para Brucelose Humana No Estado do Paraná*; Secretaria de Estado da Saúde do Paraná: Curitiba, Brazil, 2015; p. 70.
8. World Health Organization (WHO). *Brucellosis in Humans and Animals*; World Health Organization: Geneva, Switzerland, 2006; pp. 1–102.
9. Solera, J. Update on brucellosis: Therapeutic challenges. *Int. J. Antimicrob. Agents* **2010**, *36* (Suppl. 1), S18–S20. [CrossRef]
10. Young, E.J. Brucellosis: A model zoonosis in developing countries. *Apmis. Suppl.* **1988**, *3*, 17–20.
11. Dornand, J.; Gross, A.; Lafont, V.; Liautard, J.; Oliaro, J.; Liautard, J.P. The innate immune response against Brucella in humans. *Vet. Microbiol.* **2002**, *90*, 383–394. [CrossRef]
12. Campos, M.A.; Rosinha, G.M.; Almeida, I.C.; Salgueiro, X.S.; Jarvis, B.W.; Splitter, G.A.; Qureshi, N.; Bruna-Romero, O.; Gazzinelli, R.T.; Oliveira, S.C. Role of Toll-like receptor 4 in induction of cell-mediated immunity and resistance to *Brucella abortus* infection in mice. *Infect. Immun.* **2004**, *72*, 176–186. [CrossRef]
13. Huang, L.Y.; Ishii, K.J.; Akira, S.; Aliberti, J.; Golding, B. Th1-like cytokine induction by heat-killed *Brucella abortus* is dependent on triggering of TLR9. *J. Immunol.* **2005**, *175*, 3964–3970. [CrossRef] [PubMed]
14. Weiss, D.S.; Takeda, K.; Akira, S.; Zychlinsky, A.; Moreno, E. MyD88, but not toll-like receptors 4 and 2, is required for efficient clearance of *Brucella abortus*. *Infect. Immun.* **2005**, *73*, 5137–5143. [CrossRef] [PubMed]
15. Macedo, G.C.; Magnani, D.M.; Carvalho, N.B.; Bruna-Romero, O.; Gazzinelli, R.T.; Oliveira, S.C. Central role of MyD88-dependent dendritic cell maturation and proinflammatory cytokine production to control *Brucella abortus* infection. *J. Immunol.* **2008**, *180*, 1080–1087. [CrossRef] [PubMed]
16. Oliveira, S.C.; De Oliveira, F.S.; Macedo, G.C.; De Almeida, L.A.; Carvalho, N.B. The role of innate immune receptors in the control of *Brucella abortus* infection: Toll-like receptors and beyond. *Microbes Infect.* **2008**, *10*, 1005–1009. [CrossRef]
17. Gomes, M.T.; Campos, P.C.; Pereira Gde, S.; Bartholomeu, D.C.; Splitter, G.; Oliveira, S.C. TLR9 is required for MAPK/NF-kappaB activation but does not cooperate with TLR2 or TLR6 to induce host resistance to *Brucella abortus*. *J. Leukoc. Biol.* **2016**, *99*, 771–780. [CrossRef]
18. Gomes, M.T.; Campos, P.C.; Oliveira, F.S.; Corsetti, P.P.; Bortoluci, K.R.; Cunha, L.D.; Zamboni, D.S.; Oliveira, S.C. Critical role of ASC inflammasomes and bacterial type IV secretion system in caspase-1 activation and host innate resistance to *Brucella abortus* infection. *J. Immunol.* **2013**, *190*, 3629–3638. [CrossRef]
19. Marim, F.M.; Franco, M.M.C.; Gomes, M.T.R.; Miraglia, M.C.; Giambartolomei, G.H.; Oliveira, S.C. The role of NLRP3 and AIM2 in inflammasome activation during *Brucella abortus* infection. *Semin. Immunopathol.* **2017**, *39*, 215–223. [CrossRef]
20. Costa Franco, M.M.; Marim, F.; Guimaraes, E.S.; Assis, N.R.G.; Cerqueira, D.M.; Alves-Silva, J.; Harms, J.; Splitter, G.; Smith, J.; Kanneganti, T.D.; et al. *Brucella abortus* Triggers a cGAS-Independent STING Pathway To Induce Host Protection That Involves Guanylate-Binding Proteins and Inflammasome Activation. *J. Immunol.* **2018**, *200*, 607–622. [CrossRef]
21. Rakoff-Nahoum, S.; Paglino, J.; Eslami-Varzaneh, F.; Edberg, S.; Medzhitov, R. Recognition of commensal microflora by toll-like receptors is required for intestinal homeostasis. *Cell* **2004**, *118*, 229–241. [CrossRef]
22. Artis, D. Epithelial-cell recognition of commensal bacteria and maintenance of immune homeostasis in the gut. *Nat. Rev. Immunol.* **2008**, *8*, 411–420. [CrossRef]
23. Barber, G.N. STING: Infection, inflammation and cancer. *Nat. Rev. Immunol.* **2015**, *15*, 760–770. [CrossRef]
24. Hodzic, Z.; Schill, E.M.; Bolock, A.M.; Good, M. IL-33 and the intestine: The good, the bad, and the inflammatory. *Cytokine* **2017**, *100*, 1–10. [CrossRef] [PubMed]
25. Yanagisawa, K.; Takagi, T.; Tsukamoto, T.; Tetsuka, T.; Tominaga, S. Presence of a novel primary response gene ST2L, encoding a product highly similar to the interleukin 1 receptor type 1. *FEBS Lett.* **1993**, *318*, 83–87. [CrossRef]
26. Schmitz, J.; Owyang, A.; Oldham, E.; Song, Y.; Murphy, E.; McClanahan, T.K.; Zurawski, G.; Moshrefi, M.; Qin, J.; Li, X.; et al. IL-33, an interleukin-1-like cytokine that signals via the IL-1 receptor-related protein ST2 and induces T helper type 2-associated cytokines. *Immunity* **2005**, *23*, 479–490. [CrossRef] [PubMed]
27. Bergers, G.; Reikerstorfer, A.; Braselmann, S.; Graninger, P.; Busslinger, M. Alternative promoter usage of the Fos-responsive gene Fit-1 generates mRNA isoforms coding for either secreted or membrane-bound proteins related to the IL-1 receptor. *EMBO J.* **1994**, *13*, 1176–1188. [CrossRef] [PubMed]

28. Iwahana, H.; Yanagisawa, K.; Ito-Kosaka, A.; Kuroiwa, K.; Tago, K.; Komatsu, N.; Katashima, R.; Itakura, M.; Tominaga, S. Different promoter usage and multiple transcription initiation sites of the interleukin-1 receptor-related human ST2 gene in UT-7 and TM12 cells. *Eur. J. Biochem.* **1999**, *264*, 397–406. [CrossRef] [PubMed]
29. Lohning, M.; Stroehmann, A.; Coyle, A.J.; Grogan, J.L.; Lin, S.; Gutierrez-Ramos, J.C.; Levinson, D.; Radbruch, A.; Kamradt, T. T1/ST2 is preferentially expressed on murine Th2 cells, independent of interleukin 4, interleukin 5, and interleukin 10, and important for Th2 effector function. *Proc. Natl. Acad. Sci. USA* **1998**, *95*, 6930–6935. [CrossRef] [PubMed]
30. Neill, D.R.; Wong, S.H.; Bellosi, A.; Flynn, R.J.; Daly, M.; Langford, T.K.A.; Bucks, C.; Kane, C.M.; Fallon, P.G.; Pannell, R.; et al. Nuocytes represent a new innate effector leukocyte that mediates type-2 immunity. *Nature* **2010**, *464*, U1367–U1369. [CrossRef]
31. Kurowska-Stolarska, M.; Stolarski, B.; Kewin, P.; Murphy, G.; Corrigan, C.J.; Ying, S.; Pitman, N.; Mirchandani, A.; Rana, B.; Van Rooijen, N.; et al. IL-33 Amplifies the Polarization of Alternatively Activated Macrophages That Contribute to Airway Inflammation. *J. Immunol.* **2009**, *183*, 6469–6477. [CrossRef]
32. Cherry, W.B.; Yoon, J.; Barternes, K.R.; Iijima, K.; Kita, H. A novel IL-1 family cytokine, IL-33, potently activates human eosinophils. *J. Allergy Clin. Immunol.* **2008**, *121*, 1484–1490. [CrossRef]
33. Suzukawa, M.; Iikura, M.; Koketsu, R.; Nagase, H.; Tamura, C.; Komiya, A.; Nakae, S.; Matsushima, K.; Ohta, K.; Yamamoto, K.; et al. An IL-1 Cytokine Member, IL-33, Induces Human Basophil Activation via Its ST2 Receptor. *J. Immunol.* **2008**, *181*, 5981–5989. [CrossRef] [PubMed]
34. Smithgall, M.D.; Comeau, M.R.; Yoon, B.R.P.; Kaufman, D.; Armitage, R.; Smith, D.E. IL-33 amplifies both T(h)1-and T(h)2-type responses through its activity on human basophils, allergen-reactive T(h)2 cells, iNKT and NK Cells. *Int. Immunol.* **2008**, *20*, 1019–1030. [CrossRef] [PubMed]
35. Martin, N.T.; Martin, M.U. Interleukin 33 is a guardian of barriers and a local alarmin. *Nat. Immunol.* **2016**, *17*, 122–131. [CrossRef] [PubMed]
36. Le, H.; Kim, W.; Kim, J.; Cho, H.R.; Kwon, B. Interleukin-33: A mediator of inflammation targeting hematopoietic stem and progenitor cells and their progenies. *Front. Immunol.* **2013**, *4*, 104. [CrossRef] [PubMed]
37. Mahapatro, M.; Foersch, S.; Hefele, M.; He, G.W.; Giner-Ventura, E.; McHedlidze, T.; Kindermann, M.; Vetrano, S.; Danese, S.; Gunther, C.; et al. Programming of Intestinal Epithelial Differentiation by IL-33 Derived from Pericryptal Fibroblasts in Response to Systemic Infection. *Cell Rep.* **2016**, *15*, 1743–1756. [CrossRef] [PubMed]
38. Liew, F.Y.; Girard, J.P.; Turnquist, H.R. Interleukin-33 in health and disease. *Nat. Rev. Immunol.* **2016**, *16*, 676–689. [CrossRef] [PubMed]
39. Hodzic, E.; Granov, N. Gigantic Thrombus of the Left Atrium in Mitral Stenosis. *Med. Arch.* **2017**, *71*, 449–452. [CrossRef]
40. Pham, O.H.; O'Donnell, H.; Al-Shamkhani, A.; Kerrinnes, T.; Tsolis, R.M.; McSorley, S.J. T cell expression of IL-18R and DR3 is essential for non-cognate stimulation of Th1 cells and optimal clearance of intracellular bacteria. *PLoS Pathog.* **2017**, *13*, e1006566. [CrossRef]
41. Brandao, A.P.; Oliveira, F.S.; Carvalho, N.B.; Vieira, L.Q.; Azevedo, V.; Macedo, G.C.; Oliveira, S.C. Host susceptibility to *Brucella abortus* infection is more pronounced in IFN-gamma knockout than IL-12/beta2-microglobulin double-deficient mice. *Clin. Dev. Immunol.* **2012**, *2012*, 589494. [CrossRef]
42. Pasquevich, K.A.; Carabajal, M.V.; Guaimas, F.F.; Bruno, L.; Roset, M.S.; Coria, L.M.; Rey Serrantes, D.A.; Comerci, D.J.; Cassataro, J. Omp19 Enables *Brucella abortus* to Evade the Antimicrobial Activity from Host's Proteolytic Defense System. *Front. Immunol.* **2019**, *10*, 1436. [CrossRef]
43. Sangari, F.J.; Seoane, A.; Rodriguez, M.C.; Aguero, J.; Garcia Lobo, J.M. Characterization of the urease operon of *Brucella abortus* and assessment of its role in virulence of the bacterium. *Infect. Immun.* **2007**, *75*, 774–780. [CrossRef] [PubMed]
44. Delpino, M.V.; Marchesini, M.I.; Estein, S.M.; Comerci, D.J.; Cassataro, J.; Fossati, C.A.; Baldi, P.C. A bile salt hydrolase of *Brucella abortus* contributes to the establishment of a successful infection through the oral route in mice. *Infect. Immun.* **2007**, *75*, 299–305. [CrossRef] [PubMed]
45. Brenchley, J.M.; Douek, D.C. Microbial translocation across the GI tract. *Annu. Rev. Immunol.* **2012**, *30*, 149–173. [CrossRef] [PubMed]

46. Lumsden, A.B.; Henderson, J.M.; Kutner, M.H. Endotoxin levels measured by a chromogenic assay in portal, hepatic and peripheral venous blood in patients with cirrhosis. *Hepatology* **1988**, *8*, 232–236. [CrossRef] [PubMed]
47. Singh, R.; Bullard, J.; Kalra, M.; Assefa, S.; Kaul, A.K.; Vonfeldt, K.; Strom, S.C.; Conrad, R.S.; Sharp, H.L.; Kaul, R. Status of bacterial colonization, Toll-like receptor expression and nuclear factor-kappa B activation in normal and diseased human livers. *Clin. Immunol.* **2011**, *138*, 41–49. [CrossRef]
48. Neutra, M.R.; Mantis, N.J.; Kraehenbuhl, J.P. Collaboration of epithelial cells with organized mucosal lymphoid tissues. *Nat. Immunol.* **2001**, *2*, 1004–1009. [CrossRef]
49. Nakato, G.; Hase, K.; Suzuki, M.; Kimura, M.; Ato, M.; Hanazato, M.; Tobiume, M.; Horiuchi, M.; Atarashi, R.; Nishida, N.; et al. Cutting Edge: *Brucella abortus* exploits a cellular prion protein on intestinal M cells as an invasive receptor. *J. Immunol.* **2012**, *189*, 1540–1544. [CrossRef]
50. Paixao, T.A.; Roux, C.M.; Den Hartigh, A.B.; Sankaran-Walters, S.; Dandekar, S.; Santos, R.L.; Tsolis, R.M. Establishment of systemic *Brucella melitensis* infection through the digestive tract requires urease, the type IV secretion system, and lipopolysaccharide O antigen. *Infect. Immun.* **2009**, *77*, 4197–4208. [CrossRef]
51. Zhao, X.; Yang, J.; Ju, Z.; Wu, J.; Wang, L.; Lin, H.; Sun, S. *Clostridium butyricum* Ameliorates Salmonella Enteritis Induced Inflammation by Enhancing and Improving Immunity of the Intestinal Epithelial Barrier at the Intestinal Mucosal Level. *Front. Microbiol.* **2020**, *11*, 299. [CrossRef]
52. Ferrero, M.C.; Fossati, C.A.; Rumbo, M.; Baldi, P.C. Brucella invasion of human intestinal epithelial cells elicits a weak proinflammatory response but a significant CCL20 secretion. *FEMS Immunol. Med. Microbiol.* **2012**, *66*, 45–57. [CrossRef]
53. Ackermann, M.R.; Cheville, N.F.; Deyoe, B.L. Bovine ileal dome lymphoepithelial cells: Endocytosis and transport of *Brucella abortus* strain 19. *Vet. Pathol.* **1988**, *25*, 28–35. [CrossRef] [PubMed]
54. Rossetti, C.A.; Drake, K.L.; Siddavatam, P.; Lawhon, S.D.; Nunes, J.E.; Gull, T.; Khare, S.; Everts, R.E.; Lewin, H.A.; Adams, L.G. Systems biology analysis of Brucella infected Peyer's patch reveals rapid invasion with modest transient perturbations of the host transcriptome. *PLoS ONE* **2013**, *8*, e81719. [CrossRef]
55. Churin, Y.; Al-Ghoul, L.; Kepp, O.; Meyer, T.F.; Birchmeier, W.; Naumann, M. *Helicobacter pylori* CagA protein targets the c-Met receptor and enhances the motogenic response. *J. Cell Biol.* **2003**, *161*, 249–255. [CrossRef] [PubMed]
56. Shao, J.; Sheng, H. Amphiregulin promotes intestinal epithelial regeneration: Roles of intestinal subepithelial myofibroblasts. *Endocrinology* **2010**, *151*, 3728–3737. [CrossRef] [PubMed]
57. Shoyab, M.; McDonald, V.L.; Bradley, J.G.; Todaro, G.J. Amphiregulin: A bifunctional growth-modulating glycoprotein produced by the phorbol 12-myristate 13-acetate-treated human breast adenocarcinoma cell line MCF-7. *Proc. Natl. Acad. Sci. USA* **1988**, *85*, 6528–6532. [CrossRef] [PubMed]
58. Ranganathan, S.; Doucet, M.; Grassel, C.L.; Delaine-Elias, B.; Zachos, N.C.; Barry, E.M. Evaluating *Shigella flexneri* Pathogenesis in the Human Enteroid Model. *Infect. Immun.* **2019**, 87. [CrossRef]
59. Rathinam, V.A.K.; Chan, F.K. Inflammasome, Inflammation, and Tissue Homeostasis. *Trends Mol. Med.* **2018**, *24*, 304–318. [CrossRef]
60. Elinav, E.; Thaiss, C.A.; Flavell, R.A. Analysis of microbiota alterations in inflammasome-deficient mice. *Methods Mol. Biol.* **2013**, *1040*, 185–194. [CrossRef]
61. Camilleri, M.; Madsen, K.; Spiller, R.; Van Meerveld, B.G.; Verne, G.N. Intestinal barrier function in health and gastrointestinal disease (vol 24, pg 503, 2012). *Neurogastroenterol. Motil.* **2012**, *24*, 976. [CrossRef]
62. Gutierrez-Jimenez, C.; Mora-Cartin, R.; Altamirano-Silva, P.; Chacon-Diaz, C.; Chaves-Olarte, E.; Moreno, E.; Barquero-Calvo, E. Neutrophils as Trojan Horse Vehicles for *Brucella abortus* Macrophage Infection. *Front. Immunol.* **2019**, *10*, 1012. [CrossRef]
63. Staurengo-Ferrari, L.; Trevelin, S.C.; Fattori, V.; Nascimento, D.C.; De Lima, K.A.; Pelayo, J.S.; Figueiredo, F.; Casagrande, R.; Fukada, S.Y.; Teixeira, M.M.; et al. Interleukin-33 Receptor (ST2) Deficiency Improves the Outcome of *Staphylococcus aureus*-Induced Septic Arthritis. *Front. Immunol.* **2018**, *9*, 962. [CrossRef] [PubMed]
64. Yan, Y.; Kolachala, V.; Dalmasso, G.; Nguyen, H.; Laroui, H.; Sitaraman, S.V.; Merlin, D. Temporal and spatial analysis of clinical and molecular parameters in dextran sodium sulfate induced colitis. *PLoS ONE* **2009**, *4*, e6073. [CrossRef] [PubMed]

65. Vieira, A.T.; Fagundes, C.T.; Alessandri, A.L.; Castor, M.G.; Guabiraba, R.; Borges, V.O.; Silveira, K.D.; Vieira, E.L.; Goncalves, J.L.; Silva, T.A.; et al. Treatment with a novel chemokine-binding protein or eosinophil lineage-ablation protects mice from experimental colitis. *Am. J. Pathol.* **2009**, *175*, 2382–2391. [CrossRef] [PubMed]
66. Weischenfeldt, J.; Porse, B. Bone Marrow-Derived Macrophages (BMM): Isolation and Applications. *Cold Spring Harb. Protoc.* **2008**, *2008*, pdb-prot5080. [CrossRef] [PubMed]

pathogens

MDPI

Article

Brucella abortus Proliferates in Decidualized and Non-Decidualized Human Endometrial Cells Inducing a Proinflammatory Response

Lucía Zavattieri [1,2,†], Mariana C. Ferrero [1,2,†], Iván M. Alonso Paiva [1,2], Agustina D. Sotelo [1,2], Andrea M. Canellada [1,2] and Pablo C. Baldi [1,2,*]

1 Facultad de Farmacia y Bioquímica, Cátedra de Inmunología, Universidad de Buenos Aires, Buenos Aires 1113, Argentina; mv.lzavattieri@gmail.com (L.Z.); ferrerom@ffyb.uba.ar (M.C.F.); ivan_alonsopaiva@yahoo.com.ar (I.M.A.P.); agustinadsotelo@gmail.com (A.D.S.); acanell@ffyb.uba.ar (A.M.C.)

2 CONICET-Universidad de Buenos Aires, Instituto de Estudios de la Inmunidad Humoral (IDEHU), Buenos Aires 1033, Argentina

* Correspondence: pablobal@ffyb.uba.ar; Tel.: +54-11-5287-4419

† These authors contributed equally to this work.

Received: 22 March 2020; Accepted: 17 April 2020; Published: 12 May 2020

Abstract: *Brucella* spp. have been associated with abortion in humans and animals. Although the mechanisms involved are not well established, it is known that placental *Brucella* infection is accompanied by inflammatory phenomena. The ability of *Brucella abortus* to infect and survive in human endometrial stromal cells (T-HESC cell line) and the cytokine response elicited were evaluated. *B. abortus* was able to infect and proliferate in both non-decidualized and decidualized T-HESC cells. Intracellular proliferation depended on the expression of a functional *virB* operon in the pathogen. *B. abortus* internalization was inhibited by cytochalasin D and to a lower extent by colchicine, but was not affected by monodansylcadaverine. The infection did not induce cytotoxicity and did not alter the decidualization status of cells. *B. abortus* infection elicited the secretion of IL-8 and MCP-1 in either decidualized or non-decidualized T-HESC, a response also induced by heat-killed *B. abortus* and outer membrane vesicles derived from this bacterium. The stimulation of T-HESC with conditioned media from *Brucella*-infected macrophages induced the production of IL-6, MCP-1 and IL-8 in a dose-dependent manner, and this effect was shown to depend on IL-1β and TNF-α. The proinflammatory responses of T-HESC to *B. abortus* and to factors produced by infected macrophages may contribute to the gestational complications of brucellosis.

Keywords: *Brucella abortus*; human endometrial cells; internalization; intracellular replication; decidualization; chemokines; macrophages

1. Introduction

Human brucellosis, a zoonotic disease mostly caused by *Brucella melitensis*, *B. suis* and *B. abortus*, affects over 500,000 people each year around the world [1]. The infection can be found in several domestic animals (cattle, sheep, goats, pigs, and dogs) and in some wild species. Transmission to humans usually occurs by contact with infected animal tissues and consumption of dairy products.

The clinical manifestations of human brucellosis are usually linked to inflammatory phenomena in the affected organs [2]. Involvement with the reproductive organs is common in animals, which frequently present problems such as abortion and perinatal death. Studies performed in animals have shown that placental *Brucella* infection is accompanied by the infiltration of inflammatory cells [3,4]. The fact that placental inflammatory responses are involved in infection-triggered abortion by several pathogens [5–7] suggests that placental inflammation may also have a role in *Brucella*-induced abortion.

Abortion due to *Brucella* infection has been also reported in humans, with an incidence that ranges from 7% to 40% according to different studies [8–10]. Among pregnant women who presented with acute brucellosis at a Saudi Arabian hospital, 43% had spontaneous abortion during the first and second trimester, and 2% in the third trimester [11]. In spite of the importance of *Brucella*-related abortion, the pathophysiology of this complication in humans is largely unknown. Recent studies have shown that *Brucella* spp. can infect and replicate in human trophoblasts, and that the infection elicits a proinflammatory response [12,13]. These trophoblastic inflammatory responses may be relevant to the pathogenesis of abortion in human brucellosis. However, the potential of other placental cell populations to contribute to an inflammatory environment during *Brucella* infection has not been explored.

For several microorganisms that reach the placenta by the hematogenous route, including *Brucella abortus*, in vivo studies in animal models have indicated that the maternal decidua is the initial site of placental colonization [14,15]. Decidualization of the endometrium, a process essential for successful implantation and maintenance of pregnancy, involves progesterone-driven morphological and biochemical changes of fibroblast-like endometrial stromal cells (ESCs) to differentiate into decidual stromal cells (DSCs). These DSCs are characterized by the secretion of prolactin, insulin growth factor-binding protein and several cytokines that act as regulators of the innate immunity [16].

Given the relevance of the decidua as the initial site of placental colonization for several hematogenously spread infections, the ability of decidual cells to respond to pathogens is especially relevant. Primary DSC and ESC cell lines have been shown to express several Toll-like receptors (TLRs) and Nod-like receptors (NLRs), and respond to pathogen-associated molecular patterns (PAMPs) with an enhanced production of matrix metalloproteinases (MMPs) and proinflammatory cytokines including MCP-1, IL-6, IL-8, IL-1β, and CCL5 (RANTES) [17]. At least for group B streptococcal infection the cytokine response of endometrial stromal cells is modulated by decidualization, so that decidualized cells produce IL-6, TNF-α, IL-10, and TGF-β while non-decidualized cells do not [18].

In addition to decidual stromal cells, the decidua also contains significant proportions of immune cells, including macrophages, natural killer cells, dendritic cells, and T cells [19]. Early pregnancy is considered to resemble an open wound which requires a strong inflammatory response, thus the first trimester is considered a proinflammatory phase, which turns to an anti-inflammatory phase in the second trimester [20,21]. Although decidual macrophages exhibit an M2 phenotype and exert an immunosuppressive effect on local lymphocyte populations, in the context of local infection they may increase their production of proinflammatory cytokines and contribute to pregnancy disorders [19]. Of note, DSC or ESC have been shown to interact with macrophages in several ways [22,23]. In response to stimulation with lipopolysaccharide (LPS) from *Escherichia coli*, a coculture of ESC and PMA-differentiated THP-1 cells (human monocytes) produced enhanced levels of many cytokines (IL-1β, RANTES, MCP-1, IL-10, TGF-β, MIC-1, G-CSF) as compared to the respective monocultures [24]. Importantly, *B. abortus* is known to survive and replicate in macrophages from several animal species, inducing the secretion of proinflammatory cytokines [25–27].

The T-HESC cell line, derived from normal primary human ESC by telomerase immortalization, has been widely used to study several aspects of human ESC biology, including infection and cytokine production [23,24,28–31]. T-HESC are karyotypically, morphologically, and phenotypically similar to the primary parent cells, and after treatment with estradiol and medroxyprogesterone acetate (MPA) display the morphological and biochemical pattern of decidualization [32]. In the present study we evaluated the ability of *Brucella* spp. to infect and survive in decidualized T-HESC, and also assessed the cytokine production induced in these cells by the infection or by their interaction with infected macrophages.

2. Results

2.1. *Brucella abortus* Infects and Replicates in Both Decidualized and Non-Decidualized T-HESC Cells

Both decidualized and non-decidualized T-HESC endometrial cells were infected with *B. abortus* at a multiplicity of infection (MOI) of 250 bacteria/cell, and colony-forming units (CFU) of intracellular bacteria were determined at different times post-infection (p.i.). As shown in Figure 1, *B. abortus* was able to infect T-HESC cells in both conditions, although the initial number of intracellular bacteria (2 h p.i.) was slightly higher for non-decidualized cells (1125 ± 250 vs. 345 ± 32 CFU/well, mean ± SD). Besides wild type *B. abortus*, two additional strains carrying mutations in genes relevant for virulence were also tested for their capacity to infect and survive in T-HESC cells. These included a mutant lacking the *virB10* gene, widely reported as essential for the intracellular survival and replication of *Brucella* [33,34], and a double mutant lacking *btpA* and *btpB* genes which encode proteins able to interfere with TLR signaling [35,36]. As shown in Figure 1A, both mutant strains were able to infect decidualized and non-decidualized T-HESC at levels similar to the wild type strain. However, the ability to survive and replicate intracellularly differed between the *virB10* mutant and the other two strains. While CFU of intracellular bacteria increased along time for wild type *B. abortus* and the *btpAbtpB* mutant, showing intracellular replication, the CFU of the *virB10* mutant declined at the same time and no viable bacteria were detected in either condition at 48 h p.i. This later result confirmed in endometrial cells the essential role of *virB10* for the intracellular survival of *Brucella*.

Infection experiments were also carried out in the presence of specific inhibitors to examine whether *B. abortus* internalization by T-HESC cells depends on actin polymerization (cytochalasin D), microtubules (colchicine), or clathrin-mediated endocytosis (monodansylcadaverine, MDC). As shown in Figure 1B, *B. abortus* internalization was highly inhibited by cytochalasin D and to a lower extent by colchicine, but was not affected by MDC.

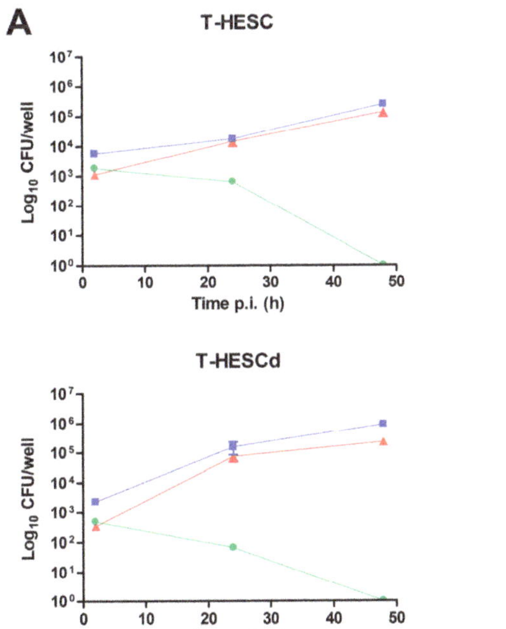

Figure 1. *Cont.*

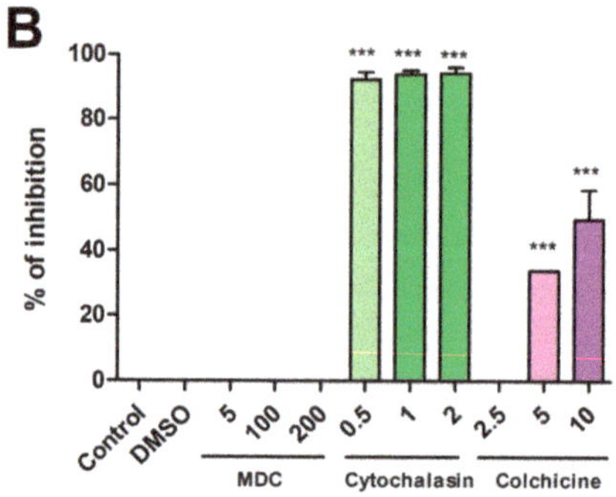

Figure 1. *Brucella abortus* invades and replicates in T-HESC cells. (**A**) Non-decidualized (T-HESC) and decidualized (T-HESCd) endometrial cells were infected with wild type *B. abortus* and two isogenic mutants (*virB10* and *btpAbtpB*), and colony forming unit (CFU) numbers of intracellular bacteria were determined at different times post-infection (p.i.). (**B**). Decidualized T-HESC were pretreated for 1 h with different doses of Colchicine (10, 5, 2.5 µM), Monodansylcadaverine (MDC; 200, 100, 5 µM), Cytochalasin D (2, 1, 0.5 µg/mL), or DMSO (vehicle) before infection with wild type *B. abortus*. Intracellular CFU were determined at 1 h p.i. Results are expressed as mean ± SD from three independent experiments run in duplicates. *** $p < 0.001$ versus control.

To assess whether the infection affected the viability of T-HESC cells or their decidualization status, the levels of lactate dehydrogenase (LDH) and prolactin were measured in culture supernatants of infected cells at 24 and 48 h p.i. and also in non-infected cells cultured in parallel. As shown in Figure 2, the infection with either wild type *B. abortus* or the *btpAbtpB* mutant did not modify the levels of LDH or prolactin as compared to non-infected cells at any time point, showing that it does not induce cytotoxicity or affect the decidualization of cells.

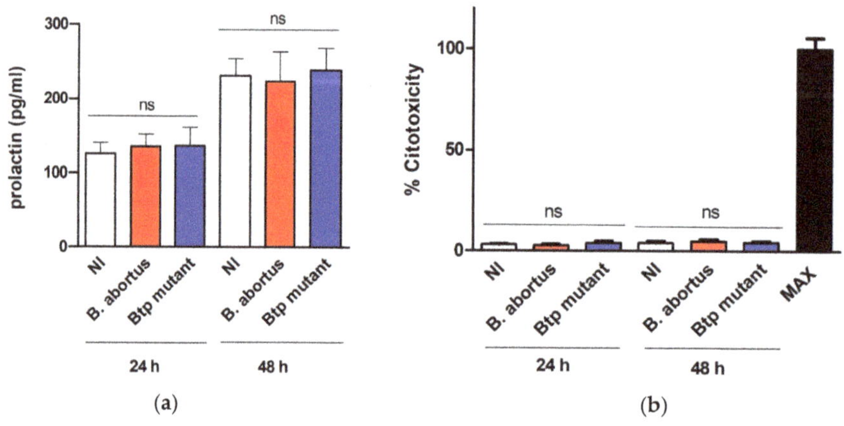

Figure 2. *B. abortus* infection does not induce cytotoxicity or alterations in decidualization in T-HESC cells. Decidualized T-HESC cells were infected or not (NI) with *B. abortus* wild type or its isogenic *btpAbtpB* mutant, and culture supernatants were harvested at 24 and 48 h p.i. to measure the levels of prolactin by ELISA (**a**) and the activity of lactate dehydrogenase (LDH) using a commercial non-radioactive cytotoxicity assay (**b**). In the latter assay, a control of 100% cell lysis (Max) was obtained by hypotonic lysis of the same number of non-infected cells. Results are expressed as mean ± SD from three independent experiments run in duplicates. ns: non-significant versus NI.

2.2. *B. abortus* Infection Induces the Secretion of Proinflammatory Chemokines in T-HESC Cells

As mentioned above, DSC and ESC cell lines express several TLRs and NLRs, and respond to microbial PAMPs with an enhanced production of proinflammatory cytokines, including MCP-1, IL-6, IL-8, IL-1β, and RANTES [17]. To assess the ability of *B. abortus* to induce a proinflammatory response in T-HESC, these cells were infected with the wild type strain and the *btpAbtpB* mutant, and the levels of IL-8 and MCP-1 were measured in culture supernatants. The studies were performed on decidualized and non-decidualized cells to determine whether the proinflammatory response depends on the decidualization status. As shown in Figure 3, the infection with any of the *B. abortus* strains elicited the secretion of both chemokines in either decidualized or non-decidualized T-HESC, and this effect was mostly evident at 48 h p.i. At this time point, IL-8 levels were higher in non-decidualized cells than in decidualized ones (mean, 8539 vs. 4948 pg/mL), whereas no significant difference was found for MCP-1 (mean, 3197 vs. 3621 pg/mL).

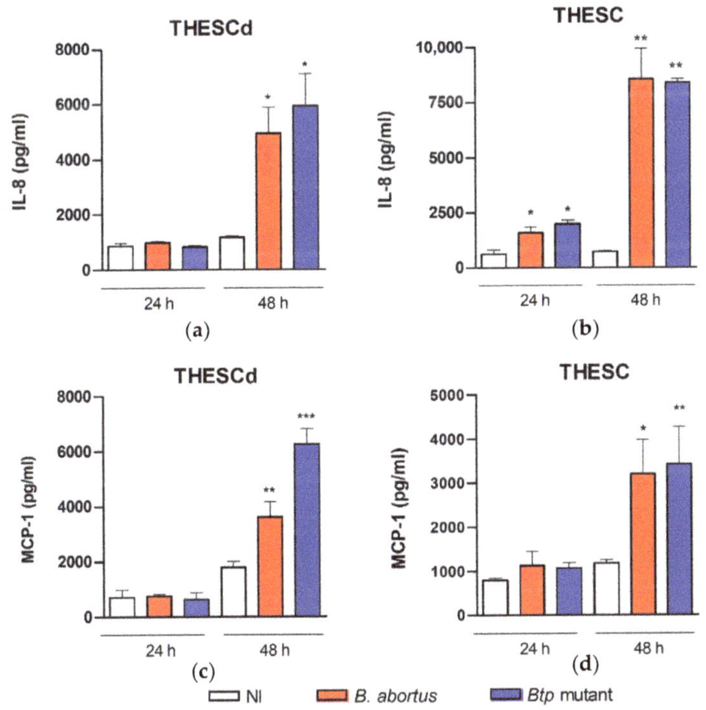

Figure 3. *B. abortus* infection elicits chemokine secretion in T-HESC cells. Decidualized (T-HESCd) (**a**,**c**) and non-decidualized (T-HESC) (**b**,**d**) endometrial cells were infected or not (NI) with wild type *B. abortus* and the *btpAbtpB* mutant, and the levels of IL-8 (**a**,**b**) and MCP-1 (**c**,**d**) were measured by ELISA in culture supernatants harvested at 24 or 48 h p.i. Results are expressed as mean ± SD from three independent experiments run in duplicates. * $p < 0.05$, ** $p < 0.01$, *** $p < 0.001$ versus NI.

To determine which signaling pathways may be involved in the induction of chemokine secretion, decidualized T-HESC cells were treated with SB203580 (p38 MAPK inhibitor), SP600125 (Jnk1/2 inhibitor), BAY 11-7082 (NF-κB inhibitor), or vehicle (dimethyl sulfoxide, DMSO) before and during the infection with *B. abortus*, and IL-8 and MCP-1 were measured as above. As shown in Figure 4, the secretion of both cytokines was not affected significantly by DMSO but was reduced to basal levels by all the inhibitors tested, suggesting that all the signaling pathways (p38, Jnk1/2, and NF-kB) are involved.

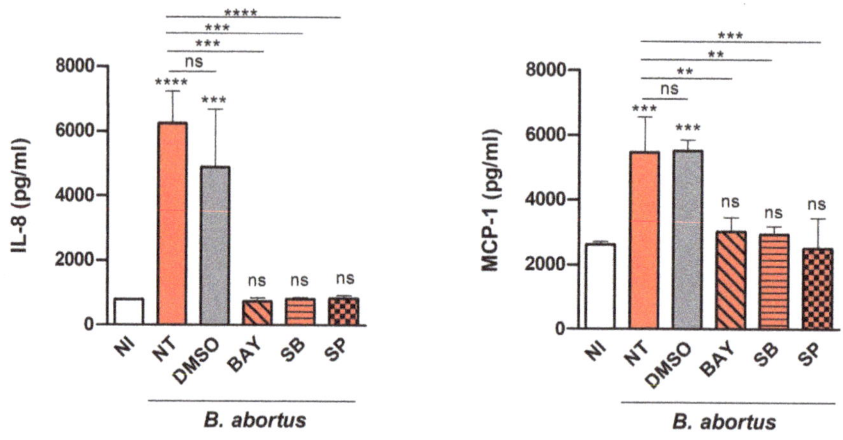

(a) (b)

Figure 4. Several signaling pathways are involved in chemokine secretion in *Brucella*-infected T-HESC cells. Decidualized T-HESC cells were treated or not (NT) with SB203580 (SB, p38 MAPK inhibitor), SP600125 (SP, Jnk1/2 inhibitor), BAY 11-7082 (BAY, NF-κB inhibitor), or vehicle (DMSO) for 1 h, and were infected with wild type *B. abortus*. The inhibitors were kept throughout the experiment. At 48 h p.i. culture supernatants were harvested for measuring IL-8 (**a**) and MCP-1 (**b**) by ELISA. Non-treated non-infected cells (NI) served as controls. Results are expressed as mean ± SD from three independent experiments run in duplicates. Asterisks over bars indicate *** $p < 0.001$ or **** $p < 0.0001$ versus NI. Asterisks over lines indicate ** $p < 0.01$, *** $p < 0.001$ or **** $p < 0.0001$ versus NT. ns: non-significant.

Given the ability of *B. abortus* infection to induce the secretion of IL-8 and MCP-1 in T-HESC cells, experiments were carried out to determine whether such responses can be also elicited by stimulation with *B. abortus* antigens or, conversely, depend on bacterial viability. For this purpose, cells were stimulated with either heat-killed *B. abortus* (HKBA), or LPS or outer membrane vesicles (OMVs) from this bacterium, and chemokine levels were measured at 48 h p.i. As shown in Figure 5, HKBA (at 10^9 CFU/mL) elicited IL-8 and MCP-1 secretion by T-HESC cells, albeit at lower levels than those attained by the infection. In addition, IL-8 secretion was significantly induced by *B. abortus* OMVs. These results show that the induction of chemokines in these cells does not depend on *Brucella* viability.

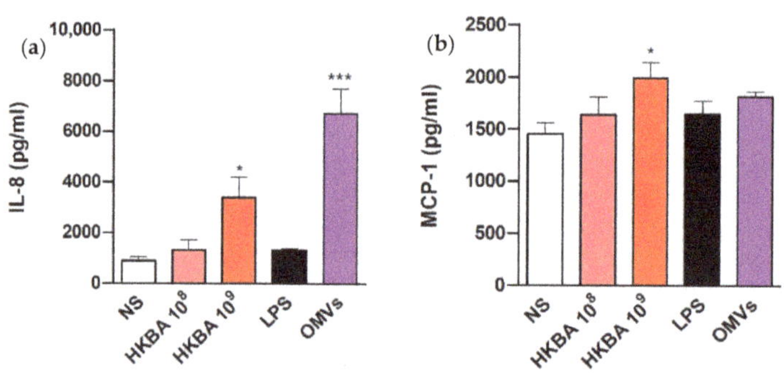

Figure 5. The chemokine response of T-HESC to *B. abortus* does not require bacterial viability. Decidualized T-HESC cells were stimulated or not (NS) with two doses (10^8 and 10^9 CFU/mL) of heat-killed *B. abortus* (HKBA), or with lipopolysaccharide (LPS) or outer membrane vesicles (OMVs) from this bacterium, and IL-8 (**a**) and MCP-1 (**b**) levels were measured in culture supernatants at 48 h post-stimulation. Results are expressed as mean ± SD from three independent experiments run in duplicates. * $p < 0.05$, *** $p < 0.001$ versus NS.

2.3. Factors Produced by Brucella-Infected Macrophages Stimulate Proinflammatory Responses in Decidualized T-HESC Cells

The results shown above demonstrate that decidualized T-HESC produce proinflammatory mediators in response to infection with *B. abortus* or stimulation with its antigens. In the context of infection in the pregnant uterus, however, endometrial cells may also receive stimulation by factors produced by adjacent infected macrophages [22,23]. To model this scenario in vitro, decidualized T-HESC cells were stimulated with conditioned media (CM) from *B. abortus*-infected macrophages and the levels of proinflammatory cytokines were measured in culture supernatants 24 h later. The preexisting levels of these cytokines in the CM were subtracted in order to calculate the secretion specifically induced by the stimulation. As shown in Figure 6, stimulation with CM from *Brucella*-infected macrophages induced the production of IL-6, MCP-1, and IL-8 in a dose-dependent manner (higher secretion for stimulation with CM diluted at 1/2). No significant secretion of any of these cytokines was induced by stimulation with CM from non-infected monocytes. Previous similar studies on the stimulation of other non-phagocytic cells have shown that IL-1β and TNF-α are involved in the inducing effect of CM from *Brucella*-infected macrophages. To test whether these cytokines are also involved in the stimulation of IL-6, IL-8, and MCP-1 in decidualized T-HESC cells, experiments were performed in which CM were preincubated with a TNF-neutralizing antibody or T-HESC were preincubated with the natural antagonist of the IL-1 receptor (IL-1Ra). As shown in Figure 6, the stimulating effect of the CM on the secretion of IL-6 was significantly reduced by both pretreatments, implying that both TNF-α and IL-1β are involved. For MCP-1 and IL-8, in contrast, only the preincubation with the anti-TNF antibody produced a significant reduction. Although the isotype control also produced a significant reduction of MCP-1 levels, the reducing effect of the specific anti-TNF antibody was much more pronounced. In summary, TNF-α and/or IL-1β are involved in the ability of CM from *Brucella*-infected macrophages to stimulate the production of proinflammatory cytokines by decidualized T-HESC.

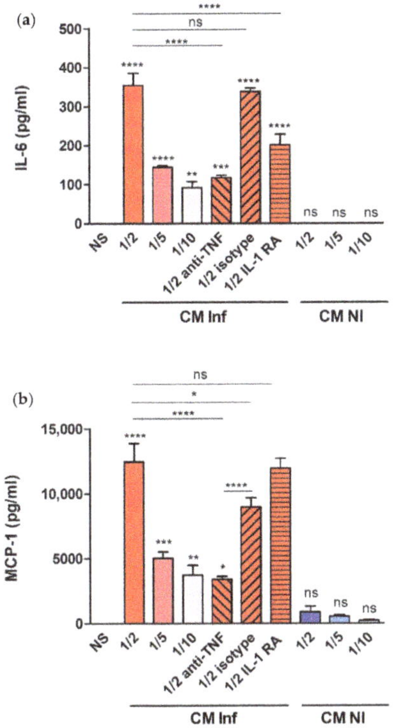

Figure 6. *Cont.*

(c)

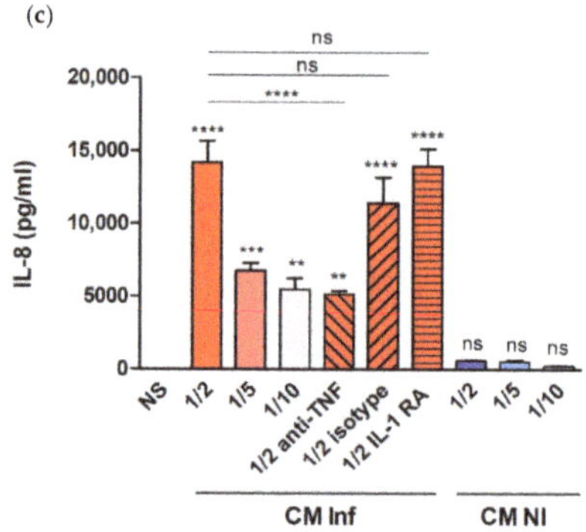

Figure 6. Factors produced by *B. abortus*–infected macrophages stimulate cytokine production by endometrial cells. Decidualized T-HESC cells were stimulated or not (NS) with conditioned media from *B. abortus*-infected macrophages (CM Inf) or from uninfected macrophages (CM NI) at different dilutions (1/2, 1/5 or 1/10), and 24 h later culture supernatants were harvested to measure IL-6 (**a**), MCP-1 (**b**) and IL-8 (**c**) levels. In parallel experiments, CM Inf was preincubated with a TNF-α neutralizing antibody or an isotype control before addition to cells, or T-HESC were preincubated with the natural antagonist of the IL-1 receptor (IL-1Ra) before stimulation with CM Inf, and cytokine levels were measured as described. Results are expressed as mean ± SD from three independent experiments run in duplicates. * $p < 0.05$, ** $p < 0.01$, *** $p < 0.001$, **** $p < 0.0001$, ns: non-significant. Asterisks over bars indicate differences versus NS.

3. Discussion

Brucella infections have been associated with abortion in both humans and animals. Although the pathophysiology of this complication has not been completely elucidated, the inflammatory phenomena observed in the affected placenta [3,4] suggest that, as with other pathogens causing abortion, placental inflammation may have a role in *Brucella*-induced abortion. As *Brucella* can reach the placenta by the hematogenous route, the maternal decidua is probably the initial site of placental colonization [14,15]. Given the known ability of decidual cells to respond to microbial PAMPs with an enhanced production of proinflammatory cytokines, and the known deleterious effect of placental inflammation on gestation, we decided to assess the ability of *Brucella* spp. to colonize decidualized stromal endometrial cells (T-HESC) and to induce the production of proinflammatory cytokines.

As shown here, *B. abortus* was able to infect both decidualized and non-decidualized T-HESC cells, although the initial number of intracellular bacteria was slightly higher for non-decidualized cells. This may relate to the fact that decidualized cells form an organized layer thus exposing less membrane surface to the environment. In addition, the pathogen was able to survive and replicate inside these cells. These findings are in line with the reported ability of *B. abortus* for intracellular replication in several phagocytic and non-phagocytic cells, including macrophages, epithelial cells and trophoblasts [13,25,37]. It has been widely demonstrated that this ability for intracellular survival in different cell types depends on the expression of a type IV secretion system encoded by the *virB* operon, which allows *Brucella* to modulate phagosome-lysosome fusion [33,34]. In line with this, we found that a *B. abortus* mutant lacking the *virB10* gene was unable to survive and replicate inside decidualized and non-decidualized T-HESC despite a similar ability of invasion compared to the wild type strain. In contrast, a mutant lacking the genes for the BtpA and BtpB proteins that interfere with TLR signaling exhibited invasion and replication abilities similar to the wild type strain. Importantly, *B. abortus* infection did not induce cytotoxicity, nor did it affect the decidualization status of cells, suggesting that the decidua might sustain the infection in affected individuals.

The mechanisms for *Brucella* invasion of non-phagocytic cells may vary according to the cell type considered. Whereas actin polymerization and microtubules have been involved in many cells [37], internalization in Vero cells does not depend on microtubules but depends on clathrin-mediated endocytosis [38]. The requirements for invasion of endometrial cells have not been reported. We found that *B. abortus* internalization was inhibited by cytochalasin D and to a lower extent by colchicine, which inhibit actin polymerization and microtubule formation, respectively. In contrast, internalization was not affected by MDC, an inhibitor of clathrin-mediated endocytosis.

As mentioned previously, placental *Brucella* infection is accompanied by the infiltration of inflammatory cells [3,4], which suggests that placental inflammation may have a role in *Brucella*-induced abortion as it does in abortion triggered by other pathogens. Our results show that *B. abortus* infection elicits the secretion of IL-8 and MCP-1 in either decidualized or non-decidualized T-HESC cells. IL-8 levels were higher in non-decidualized cells than in decidualized ones, whereas no significant difference was found for MCP-1. The higher production of IL-8 in non-decidualized cells may relate to the higher number of intracellular bacteria found in this condition as compared to decidualized cells, a downmodulating effect of decidualization on IL-8 production [39], or both. Nonetheless, these results suggest that, although the decidualization status may influence the levels of some proinflammatory mediators, decidualized endometrial cells are capable of mediating a proinflammatory response to *B. abortus*. A few previous studies have shown that *Brucella* BtpA and BtpB proteins, which contain TIR motifs and can thus modulate TLR signaling, can reduce cytokine production in dendritic cells in vitro (IL-12, TNF-α) and in lung tissues in vivo (IL-12, CXCL-1, MCP-1) [35,36]. However, the potential modulating role of these proteins in *Brucella*-infected endometrial cells was unknown. At variance with those previous studies, we did not detect significant differences in the levels of the two chemokines here evaluated (IL-8 and MCP-1) between T-HESC infected with the wild type *B. abortus* strain or the *btpAbtpB* mutant strain. These results agree with those reported for the same chemokines in *Brucella*-infected human trophoblasts [12], and add support to the hypothesis that the immune responses of professional phagocytes are more influenced by the action of Btp proteins than those of non-phagocytic cells.

The secretion of both cytokines was reduced to basal levels by all the inhibitors tested, suggesting that all the signaling pathways (p38, Jnk1/2, and NF-kB) are involved. In line with these findings, previous studies have shown that several signaling pathways are involved in cytokine production by different cell types in response to *B. abortus*. For example, CCL20 secretion by human bronchial epithelial cells depends on p38, Jnk1/2, Erk1/2, and NF-kB [40], whereas in murine astrocytes TNF-α secretion depends on p38 and Erk1/2 signaling pathways [41].

Previous studies in several non-phagocytic cells have shown that not only live *B. abortus* but also some of its antigens can elicit the production of proinflammatory cytokines [42–44]. In line with these reports, we found that HKBA and OMVs from *B. abortus* elicit IL-8 and/or MCP-1 secretion in T-HESC cells. Obviously, these findings imply that the induction of chemokines in these cells does not depend on *Brucella* viability. At variance with HKBA and OMVs, *B. abortus* LPS did not elicit the production of the chemokines analyzed. This result is in line with previous studies in other cell types, which demonstrated that *B. abortus* LPS is a poor inducer of proinflammatory responses [42–45]. In contrast, most inflammatory responses are triggered by outer membrane lipoproteins, which induce TLR2 signaling [45].

As shown in this study, decidualized T-HESC produce proinflammatory mediators, including MCP-1, in response to infection with *B. abortus* or stimulation with its antigens. However, the decidua contains not only DSC but also a significant proportion of macrophages [19], with which DSC can establish reciprocal interactions [22,23]. In addition, the number of decidual macrophages could eventually augment in the context of locally increased MCP-1 levels induced by an infectious process. Therefore, it can be speculated that, during *B. abortus* infection in the pregnant uterus, endometrial cells may respond not only to the stimulus of bacterial antigens but also to stimulation by factors produced by adjacent *Brucella*-infected macrophages. In support of this hypothesis, we found that the stimulation of decidualized T-HESC with CM from *B. abortus*-infected macrophages induced the production of IL-6, MCP-1, and IL-8 in a dose-dependent manner, a phenomenon not produced by stimulation with CM from non-infected monocytes. Additional studies using specific blocking agents revealed that IL-6 induction by CM is mediated by TNF-α and IL-1β, whereas the induction of MCP-1 and IL-8 is mediated by TNF-α. These findings are similar to those reported for the interaction between *Brucella*-infected macrophages and human trophoblasts [12].

Overall, these results suggest a possible scenario in which DSC produce IL-6 and chemoattractants for monocytes/macrophages in response to *B. abortus* infection and/or in response to cytokines produced by *Brucella*-infected placental macrophages. Reciprocal stimulations between DSC and phagocytes may amplify these phenomena. These interactions may be long-lasting due to the ability of *Brucella* to survive and replicate within macrophages and DSC. Altogether, these proinflammatory responses may contribute to the gestational complications of brucellosis.

4. Materials and Methods

4.1. Reagents

LPS from *Brucella abortus* 2308 was provided by Ignacio Moriyón (University of Navarra, Pamplona, Spain). The purity and the characteristics of this preparation have been published previously [46].

4.2. Cell lines

A human endometrial stromal cell line (T-HESC) was kindly provided Dr. Andrea Randi (Human Biochemistry Department, School of Medicine, University of Buenos Aires). This cell line was derived from normal stromal cells obtained from an adult patient subjected to hysterectomy, and conserved the characteristics of the regular endometrial cells [32]. The line was obtained by immortalization by transfection of telomerase (hTERT) using a retroviral system, and expressed puromycin resistance genes. Cells were maintained in DMEM-F12 supplemented with 10% FCS, 50 U/mL penicillin, 50 μg/mL streptomycin, 2 mM glutamine and 500 ng/mL puromycin. Decidualization was achieved following published procedures [47]. Briefly, T-HESC (5×10^4 cells/well) were treated with medroxyprogesterone acetate (MPA, 10^{-7} M) and dibutyryl cAMP (0.5 mM) for 8 days, changing the culture media every 48 h. Decidualization was evaluated by morphology and by prolactin levels measured by sandwich ELISA (R&D Systems). For infection assays, cells were cultured for 24 h in antibiotic-free culture medium.

4.3. Monocyte Isolation and Macrophage Differentiation

Peripheral blood samples were obtained from healthy volunteers after approval by the Ethics Committee of the School of Pharmacy and Biochemistry (Approval 2194/17). Written informed consent was obtained from all volunteers. Human monocytes were isolated by standard procedures. Briefly, whole blood diluted with sterile phosphate-buffered saline (PBS) was carefully layered on Ficoll-Paque (density: 1.077 g/mL) and centrifuged at 400× *g* for 30 min. The layer containing peripheral blood mononuclear cells was carefully removed by pipetting and washed with PBS by centrifugation at 250× *g* for 10 min. The pellet was resuspended in RPMI 1640 medium supplemented with 1 mM glutamine and was incubated for 2 h in 24-well plates. After washing with sterile PBS to eliminate nonadherent cells, RPMI medium supplemented with 10% sera from the same donors and antibiotics (100 U/mL penicillin and 100 μg/mL streptomycin) was added to the adherent cells. Cells were incubated at 37 °C in a 5% CO_2 atmosphere for 7 days for in vitro macrophagic differentiation [48]. Antibiotics were removed 24 h prior to infection.

4.4. Bacterial Strains and Growth Conditions

B. abortus 2308 (wild type strain), its isogenic *btpAbtpB* double mutant and *virB10* polar mutant were obtained from our collection. The strains were grown in tryptic soy broth at 37 °C with agitation. After two washes with sterile PBS, bacterial inocula were adjusted to the desired concentration in sterile PBS based on optical density readings. An aliquot of each suspension was plated on tryptic soy agar (TSA) and incubated at 37 °C to determine the actual concentration of colony-forming units

(CFU) in the inocula. Cells were inoculated with *B. abortus* 2308 at an MOI of 200 and the plates were centrifuged for 10 min at 1200 rpm at room temperature. After 2 h, culture medium was removed and replaced with medium containing gentamicin and streptomycin. All live *Brucella* manipulations were performed in biosafety level 3 facilities. To prepare HKBA, bacteria were washed in sterile PBS, heat killed at 70 °C for 30 min, aliquoted, and stored at −80 °C until use. The absence of bacterial viability was checked by plating on TSA.

4.5. Isolation of Outer Membrane Vesicles

OMVs from *B. abortus* 2308 were obtained essentially as described previously [49]. Briefly, bacteria were grown as described above, harvested by centrifugation and washed twice in sterile PBS. The pellet was resuspended in Gerhardt-Wilson minimal medium at an $OD_{600\ nm}$ of 0.1 and cultured for 72 h (early stationary phase of growth). The culture was centrifuged, and the cell-free supernatant was filter-sterilized. The filtrate was centrifuged at 15,000× *g* for 5 h at 4 °C. The pellets containing the OMVs were resuspended in PBS, and protein concentration was measured by the bicinchoninic acid assay (Pierce). The presence of OMVs was corroborated by electron microscopy. OMVs were stored at −20 °C until use.

4.6. Stimulation of T-HESC Cells with Brucella antigens

Decidualized T-HESC cells (5×10^4 cells/well) were stimulated with LPS from *B. abortus* (1 μg/mL), OMVs (1 μg/mL of protein), or HKBA (10^9 or 10^8 CFU/mL). Cells were cultured at 37 °C in a 5% CO_2 atmosphere, and supernatants were collected 48 h after stimulation for chemokine measurement.

4.7. Cellular Infections

Decidualized and non-decidualized T-HESC cells were infected with *B. abortus* 2308 at MOI of 250 bacteria/cell. Monocyte-derived macrophages were infected at MOI 100 bacteria/cell in culture medium containing no antibiotics. After dispensing the bacterial suspension, the plates were centrifuged (10 min at 400× *g*) and then incubated for 2 h at 37 °C in a 5% CO_2 atmosphere. Non-internalized bacteria were eliminated by several washes with medium alone followed by incubation in medium supplemented with 100 μg/mL gentamicin and 50 μg/mL streptomycin. After that, cells were washed and then incubated with culture medium without antibiotics. At different times post-infection (2, 24 or 48 h) culture supernatants were harvested for cytokine measurement, while the cells were washed with sterile PBS and lysed with 0.2% Triton X-100. Serial dilutions of the lysates were plated on TSA to enumerate intracellular CFU. In addition, the levels of prolactin were measured in culture supernatants as described above to assess the impact of infection on the decidualization status of the cells, and the levels of LDH were measured to assess cytotoxicity.

4.8. Evaluation of Cytotoxicity

To analyze the effect of infection on cell integrity, the release of lactate dehydrogenase (LDH) from infected T-HESC cells was determined. LDH activity was determined using the CytotTox 96 Non-Radiactive Cytotoxicity Assay (Promega, USA) in culture supernatants obtained at 24 and 48 h p.i. Results were expressed as the ratio between LDH levels measured in the samples (infected or non-infected cultures) and those corresponding to a 100% cell lysis (obtained by hypotonic lysis of the same number of cells).

4.9. Internalization Pathways

To assess the role of microtubules, actin or clathrin in *B. abortus* internalization, decidualized T-HESC were pretreated for 1 h with different doses of colchicine (10, 5, 2.5 μM, Sigma), monodansylcadaverine (MDC; 200, 100, 5 μM) or cytochalasin D (2, 1, 0.5 μg/mL, Sigma) and were later infected as described above but in the presence of these inhibitors. MDC and cytochalasin were solubilized in dimethyl sulfoxide (DMSO), and in all the experiments control cells were incubated without inhibitor or with DMSO for the same period as treated cells. Intracellular CFU were determined at 2 h p.i. as described above.

4.10. Stimulation of T-HESC with Conditioned Media (CM) from Brucella-Infected Macrophages

CM from macrophages infected with *B. abortus* 2308 (MOI 100) were harvested at 24 h p.i., filter-sterilized and used to stimulate noninfected decidualized T-HESC cells. After 24 and 48 h, supernatants from stimulated cultures were harvested to measure cytokines. The preexisting levels of cytokines in the CM were subtracted in order to calculate the secretion specifically induced by the stimulation. To determine if TNF-α might be involved in the stimulating effects of CM, in some experiments CM were preincubated for 1 h at 37 °C with a neutralizing monoclonal antibody against TNF-α or its isotype control (both from BD Pharmingen) before being transferred to T-HESC cultures. Alternatively, to determine the role of IL-1 β in the stimulating effect, decidualized T-HESC cells were preincubated with the IL-1β receptor antagonist IL-1Ra (R&D Systems) for 1 h at 37 °C before stimulation with CM from infected macrophages.

4.11. Inhibition of Signaling Pathways

To examine the signaling pathways involved in cytokine secretion, decidualized T-HESC cells were pretreated with 10μM SB203580 (p38 MAPK inhibitor, Gibco), 10 μM SP600125 (Jnk1/2 inhibitor, Sigma), 2.5 μM BAY11-7082 (NF-κB inhibitor, Sigma) or vehicle (DMSO). These reagents were added one hour before infection with *B. abortus* and were kept throughout the experiment (48 h). Cell viability after incubation with these inhibitors was higher than 90%, as assessed by staining with trypan blue.

4.12. Measurement of Cytokines and Chemokines

Levels of human IL-6, IL-8, MCP-1, and TNF-α were measured in culture supernatants by sandwich ELISA, using paired cytokine-specific monoclonal antibodies according to the manufacturer's instructions (BD Pharmingen).

4.13. Statistical Analysis

Each experiment was performed in duplicates on three independent occasions. The values obtained are presented as the mean ± SD. Statistical analysis was performed with one-way ANOVA, followed by Post Hoc Tukey's Test or Dunnett's Test using GraphPad Prism 6.0 software.

Author Contributions: Conceptualization, M.C.F., A.M.C., and P.C.B.; Methodology, L.Z., M.C.F., I.M.A.P., and A.D.S.; Formal analysis, L.Z., M.C.F., A.M.C., and P.C.B.; Writing—original draft preparation, L.Z., M.C.F., I.M.A.P., A.D.S., A.M.C., and P.C.B.; Supervision, A.M.C. and P.C.B. All authors have read and agreed to the published version of the manuscript.

Funding: This work was supported by a Research Grant from Agencia Nacional de Promoción Científica (PICT 2016-1199), and by two Research Grants from Universidad de Buenos Aires (UBACYT 20020170100036BA and 20020160100126BA).

Acknowledgments: We thank Andrea Randi for providing the T-HESC cells. We are grateful to the staff of the BSL3 facility of INBIRS for expert assistance.

Conflicts of Interest: The authors declare no conflict of interest. The funders had no role in the design of the study; in the collection, analyses, or interpretation of data; in the writing of the manuscript, or in the decision to publish the results.

References

1. Pappas, G.; Papadimitriou, P.; Akritidis, N.; Christou, L.; Tsianos, E.V. The new global map of human brucellosis. *Lancet Infect. Dis.* **2006**, *6*, 91–99. [CrossRef]
2. Pappas, G.; Akritidis, N.; Bosilkovski, M.; Tsianos, E. Brucellosis. *N. Engl. J. Med.* **2005**, *352*, 2325–2336. [CrossRef] [PubMed]

3. Carmichael, L.E.; Kenney, R.M. Canine abortion caused by Brucella canis. *J. Am. Vet. Med. Assoc.* **1968**, *152*, 605–616.
4. Carvalho, A.V.; Mol, J.P.S.; Xavier, M.N.; Paixão, T.A.; Lage, A.P.; Santos, R.L. Pathogenesis of bovine brucellosis. *Vet. J.* **2010**, *184*, 146–155.
5. Bildfell, R.J.; Thomson, G.W.; Haines, D.M.; McEwen, B.J.; Smart, N. Coxiella burnetii infection is associated with placentitis in cases of bovine abortion. *J. Vet. Diagn. Invest.* **2000**, *12*, 419–425. [CrossRef]
6. Buxton, D.; Anderson, I.E.; Longbottom, D.; Livingstone, M.; Wattegedera, S.; Entrican, G. Ovine chlamydial abortion: Characterization of the inflammatory immune response in placental tissues. *J. Comp. Pathol.* **2002**, *127*, 133–141. [CrossRef]
7. Chattopadhyay, A.; Robinson, N.; Sandhu, J.K.; Finlay, B.B.; Sad, S.; Krishnan, L. Salmonella enterica Serovar Typhimurium-Induced Placental Inflammation and Not Bacterial Burden Correlates with Pathology and Fatal Maternal Disease. *Infect. Immun.* **2010**, *78*, 2292–2301. [CrossRef]
8. Lulu, A.R.; Araj, G.F.; Khateeb, M.I.; Mustafa, M.Y. Human Brucellosis in Kuwait: A Prospective Study of 400 Cases. *Q J. Med.* **1988**, *66*, 39–54.
9. Sarram, M.; Feiz, J.; Foruzandeh, M.; Gazanfarpour, P. Intrauterine fetal infection with Brucella melitensis as a possible cause of second-trimester abortion. *Am. J. Obstet. Gynecol.* **1974**, *119*, 657–660. [CrossRef]
10. Makhseed, M.; Harouny, A.; Araj, G.; Moussa, M.A.; Sharma, P. Obstetric and gynecologic implication of brucellosis in Kuwait. *J. Perinatol.* **1998**, *18*, 196–199.
11. Khan, M.Y.; Mah, M.W.; Memish, Z.A. Brucellosis in Pregnant Women. *Clin. Infect. Dis.* **2001**, *32*, 1172–1177. [CrossRef] [PubMed]
12. Fernández, A.G.; Ferrero, M.C.; Hielpos, M.S.; Fossati, C.A.; Baldi, P.C. Proinflammatory Response of Human Trophoblastic Cells to Brucella abortus Infection and upon Interactions with Infected Phagocytes. *Biol. Reprod.* **2016**, *94*, 48. [CrossRef] [PubMed]
13. Salcedo, S.P.; Chevrier, N.; Lacerda, T.L.S.; Ben Amara, A.; Gerart, S.; Gorvel, V.A.; De Chastellier, C.; Blasco, J.M.; Mege, J.L.; Gorvel, J.P. Pathogenic brucellae replicate in human trophoblasts. *J. Infect. Dis.* **2013**, *207*, 1075–1083. [CrossRef] [PubMed]
14. Robbins, J.R.; Bakardjiev, A.I. Pathogens and the placental fortress. *Curr. Opin. Microbiol.* **2012**, *15*, 36–43. [CrossRef]
15. Vigliani, M.B.; Bakardjiev, A.I. Intracellular organisms as placental invaders. *Fetal Matern. Med. Rev.* **2014**, *25*, 332–338. [CrossRef]
16. Dunn, C.L.; Kelly, R.W.; Critchley, H.O.D. Decidualization of the human endometrial stromal cell: An enigmatic transformation. *Reprod. Biomed. Online* **2003**, *7*, 151–161. [CrossRef]
17. Anders, A.P.; Gaddy, J.A.; Doster, R.S.; Aronoff, D.M. Current concepts in maternal-fetal immunology: Recognition and response to microbial pathogens by decidual stromal cells. *Am. J. Reprod. Immunol.* **2017**, *77*, e12623. [CrossRef]
18. Castro-Leyva, V.; Zaga-Clavellina, V.; Espejel-Nuñez, A.; Vega-Sanchez, R.; Flores-Pliego, A.; Reyes-Muñoz, E.; Giono-Cerezo, S.; Nava-Salazar, S.; Espino y Sosa, S.; Estrada-Gutierrez, G. Decidualization Mediated by Steroid Hormones Modulates the Innate Immunity in Response to Group B Streptococcal Infection in vitro. *Gynecol. Obstet. Invest.* **2017**, *82*, 592–600. [CrossRef]
19. Nagamatsu, T.; Schust, D.J. Review: The Immunomodulatory Roles of Macrophages at the Maternal-Fetal Interface. *Reprod. Sci.* **2010**, *17*, 209–218. [CrossRef]
20. Mor, G.; Cardenas, I.; Abrahams, V.; Guller, S. Inflammation and pregnancy: The role of the immune system at the implantation site. *Ann. N. Y. Acad. Sci.* **2011**, *1221*, 80–87. [CrossRef]
21. Mor, G.; Cardenas, I. The Immune System in Pregnancy: A Unique Complexity. *Am. J. Reprod. Immunol.* **2010**, *63*, 425–433. [CrossRef] [PubMed]
22. Oreshkova, T.; Dimitrov, R.; Mourdjeva, M. A Cross-Talk of Decidual Stromal Cells, Trophoblast, and Immune Cells: A Prerequisite for the Success of Pregnancy. *Am. J. Reprod. Immunol.* **2012**, *68*, 366–373. [CrossRef] [PubMed]
23. Eyster, K.M.; Hansen, K.A.; Winterton, E.; Klinkova, O.; Drappeau, D.; Mark-Kappeler, C.J. Reciprocal Communication Between Endometrial Stromal Cells and Macrophages. *Reprod. Sci.* **2010**, *17*, 809–822. [CrossRef] [PubMed]

24. Rogers, L.M.; Anders, A.P.; Doster, R.S.; Gill, E.A.; Gnecco, J.S.; Holley, J.M.; Randis, T.M.; Ratner, A.J.; Gaddy, J.A.; Osteen, K.; et al. Decidual stromal cell-derived PGE $_2$ regulates macrophage responses to microbial threat. *Am. J. Reprod. Immunol.* **2018**, *80*, e13032. [CrossRef] [PubMed]
25. Celli, J. Surviving inside a macrophage: The many ways of Brucella. *Res. Microbiol.* **2006**, *157*, 93–98. [CrossRef]
26. Zhan, Y.; Cheers, C. Differential induction of macrophage-derived cytokines by live and dead intracellular bacteria in vitro. *Infect. Immun.* **1995**, *63*, 720–723. [CrossRef]
27. Covert, J.; Mathison, A.J.; Eskra, L.; Banai, M.; Splitter, G. Brucella melitensis, B. neotomae and B. ovis elicit common and distinctive macrophage defense transcriptional responses. *Exp. Biol. Med.* **2009**, *234*, 1450–1467. [CrossRef]
28. Pagani, I.; Ghezzi, S.; Ulisse, A.; Rubio, A.; Turrini, F.; Garavaglia, E.; Candiani, M.; Castilletti, C.; Ippolito, G.; Poli, G.; et al. Human Endometrial Stromal Cells Are Highly Permissive To Productive Infection by Zika Virus. *Sci. Rep.* **2017**, *7*, 44286. [CrossRef]
29. Lu, Y.; Bocca, S.; Anderson, S.; Wang, H.; Manhua, C.; Beydoun, H.; Oehninger, S. Modulation of the Expression of the Transcription Factors T-Bet and GATA-3 in Immortalized Human Endometrial Stromal Cells (HESCs) by Sex Steroid Hormones and cAMP. *Reprod. Sci.* **2013**, *20*, 699–709. [CrossRef]
30. Gellersen, B.; Wolf, A.; Kruse, M.; Schwenke, M.; Bamberger, A.-M. Human Endometrial Stromal Cell-Trophoblast Interactions: Mutual Stimulation of Chemotactic Migration and Promigratory Roles of Cell Surface Molecules CD82 and CEACAM11. *Biol. Reprod.* **2013**, *88*, 80. [CrossRef]
31. Grasso, E.; Gori, S.; Soczewski, E.; Fernández, L.; Gallino, L.; Vota, D.; Martínez, G.; Irigoyen, M.; Ruhlmann, C.; Lobo, T.F.; et al. Impact of the Reticular Stress and Unfolded Protein Response on the inflammatory response in endometrial stromal cells. *Sci. Rep.* **2018**, *8*, 12274. [CrossRef] [PubMed]
32. Krikun, G.; Mor, G.; Alvero, A.; Guller, S.; Schatz, F.; Sapi, E.; Rahman, M.; Caze, R.; Qumsiyeh, M.; Lockwood, C.J. A Novel Immortalized Human Endometrial Stromal Cell Line with Normal Progestational Response. *Endocrinology* **2004**, *145*, 2291–2296. [CrossRef] [PubMed]
33. Sieira, R.; Comerci, D.J.; Sanchez, D.O.; Ugalde, R.A. A homologue of an operon required for DNA transfer in Agrobacterium is required in Brucella abortus for virulence and intracellular multiplication. *J. Bacteriol.* **2000**, *182*, 4849–4855. [CrossRef] [PubMed]
34. Comerci, D.J.; Martínez-Lorenzo, M.J.; Sieira, R.; Gorvel, J.P.; Ugalde, R.A. Essential role of the virB machinery in the maturation of the Brucella abortus-containing vacuole. *Cell. Microbiol.* **2001**, *3*, 159–168. [CrossRef]
35. Salcedo, S.P.; Marchesini, M.I.; Degos, C.; Terwagne, M.; Von Bargen, K.; Lepidi, H.; Herrmann, C.K.; Santos Lacerda, T.L.; Imbert, P.R.C.; Pierre, P.; et al. BtpB, a novel Brucella TIR-containing effector protein with immune modulatory functions. *Front. Cell. Infect. Microbiol.* **2013**, *3*, 28. [CrossRef]
36. Hielpos, M.S.; Ferrero, M.C.; Fernández, A.G.; Falivene, J.; Vanzulli, S.; Comerci, D.J.; Baldi, P.C. Btp Proteins from Brucella abortus modulate the lung innate immune response to infection by the respiratory route. *Front. Immunol.* **2017**, *8*, 1011. [CrossRef]
37. Pizarro-Cerdá, J.; Moreno, E.; Gorvel, J. Invasion and intracellular trafficking of Brucella abortus in nonphagocytic cells. *Microbes Infect.* **2000**, *2*, 829–835. [CrossRef]
38. Detilleux, P.G.; Deyoe, B.L.; Cheville, N.F. Effect of endocytic and metabolic inhibitors on the internalization and intracellular growth of Brucella abortus in Vero cells. *Am. J. Vet. Res.* **1991**, *52*, 1658–1664.
39. Lockwood, C.J.; Kumar, P.; Krikun, G.; Kadner, S.; Dubon, P.; Critchley, H.; Schatz, F. Effects of thrombin, hypoxia, and steroids on interleukin-8 expression in decidualized human endometrial stromal cells: Implications for long-term progestin-only contraceptive-induced bleeding. *J. Clin. Endocrinol. Metab.* **2004**, *89*, 1467–1475. [CrossRef]
40. Hielpos, M.S.; Ferrero, M.C.; Fernández, A.G.; Bonetto, J.; Giambartolomei, G.H.; Fossati, C.A.; Baldi, P.C. CCL20 and beta-defensin 2 production by human lung epithelial cells and macrophages in response to Brucella abortus infection. *PLoS ONE* **2015**, *10*, e0140408. [CrossRef]
41. Miraglia, M.C.; Scian, R.; Samartino, C.G.; Barrionuevo, P.; Rodriguez, A.M.; Ibañez, A.E.; Coria, L.M.; Velásquez, L.N.; Baldi, P.C.; Cassataro, J.; et al. Brucella abortus induces TNF-α-dependent astroglial MMP-9 secretion through mitogen-activated protein kinases. *J. Neuroinflamm.* **2013**, *10*, 47. [CrossRef] [PubMed]
42. Ferrero, M.C.; Bregante, J.; Delpino, M.V.; Barrionuevo, P.; Fossati, C.A.; Giambartolomei, G.H.; Baldi, P.C. Proinflammatory response of human endothelial cells to Brucella infection. *Microbes Infect.* **2011**, *13*, 852–861. [CrossRef] [PubMed]

43. Scian, R.; Barrionuevo, P.; Giambartolomei, G.H.; De Simone, E.A.; Vanzulli, S.I.; Fossati, C.A.; Baldi, P.C.; Delpino, M.V. Potential role of fibroblast-like synoviocytes in joint damage induced by Brucella abortus infection through production and induction of matrix metalloproteinases. *Infect. Immun.* **2011**, *79*, 3619–3632. [CrossRef] [PubMed]
44. Ferrero, M.C.; Fossati, C.A.; Baldi, P.C. Direct and monocyte-induced innate immune response of human lung epithelial cells to Brucella abortus infection. *Microbes Infect.* **2010**, *12*, 736–747. [CrossRef]
45. Giambartolomei, G.H.; Zwerdling, A.; Cassataro, J.; Bruno, L.; Fossati, C.A.; Philipp, M.T. Lipoproteins, not lipopolysaccharide, are the key mediators of the proinflammatory response elicited by heat-killed Brucella abortus. *J. Immunol.* **2004**, *173*, 4635–4642. [CrossRef]
46. Velasco, J.; Bengoechea, J.A.; Brandenburg, K.; Lindner, B.; Seydel, U.; González, D.; Zähringer, U.; Moreno, E.; Moriyón, I. Brucella abortus and its closest phylogenetic relative, Ochrobactrum spp., differ in outer membrane permeability and cationic peptide resistance. *Infect. Immun.* **2000**, *68*, 3210–3218. [CrossRef]
47. Grasso, E.; Gori, S.; Paparini, D.; Soczewski, E.; Fernández, L.; Gallino, L.; Salamone, G.; Martinez, G.; Irigoyen, M.; Ruhlmann, C.; et al. VIP induces the decidualization program and conditions the immunoregulation of the implantation process. *Mol. Cell. Endocrinol.* **2018**, *460*, 63–72. [CrossRef]
48. Daigneault, M.; Preston, J.A.; Marriott, H.M.; Whyte, M.K.B.; Dockrell, D.H. The identification of markers of macrophage differentiation in PMA-stimulated THP-1 cells and monocyte-derived macrophages. *PLoS ONE* **2010**, *5*, e8668. [CrossRef]
49. Pollak, C.N.; Delpino, M.V.; Fossati, C.A.; Baldi, P.C. Outer Membrane Vesicles from Brucella abortus Promote Bacterial Internalization by Human Monocytes and Modulate Their Innate Immune Response. *PLoS ONE* **2012**, *7*, e50214. [CrossRef]

pathogens

MDPI

Article

Hepatic Stellate Cells and Hepatocytes as Liver Antigen-Presenting Cells during *B. abortus* Infection

Paula Constanza Arriola Benitez [1], Ayelén Ivana Pesce Viglietti [1], María Mercedes Elizalde [2], Guillermo Hernán Giambartolomei [1], Jorge Fabián Quarleri [2,*] and María Victoria Delpino [1,*]

[1] Instituto de Inmunología, Genética y Metabolismo (INIGEM), Universidad de Buenos Aires, CONICET, Buenos Aires 1120, Argentina; constanza-arriola@hotmail.com (P.C.A.B.); ayelenpv@gmail.com (A.I.P.V.); ggiambart@ffyb.uba.ar (G.H.G.)

[2] Instituto de Investigaciones Biomédicas en Retrovirus y Sida (INBIRS), Universidad de Buenos Aires, CONICET, Buenos Aires 1121, Argentina; mecheeli@hotmail.com

* Correspondence: quarleri@fmed.uba.ar (J.F.Q.); mdelpino@ffyb.uba.ar (M.V.D.); Tel.: +54-11-5950-8755 (J.F.Q. & M.V.D.); Fax: 54-11-5950-8758 (J.F.Q. & M.V.D.)

Received: 27 May 2020; Accepted: 15 June 2020; Published: 30 June 2020

Abstract: In Brucellosis, the role of hepatic stellate cells (HSCs) in the induction of liver fibrosis has been elucidated recently. Here, we study how the infection modulates the antigen-presenting capacity of LX-2 cells. *Brucella abortus* infection induces the upregulation of class II transactivator protein (CIITA) with concomitant MHC-I and -II expression in LX-2 cells in a manner that is independent from the expression of the type 4 secretion system (T4SS). In concordance, *B. abortus* infection increases the phagocytic ability of LX-2 cells and induces MHC-II-restricted antigen processing and presentation. In view of the ability of *B. abortus*-infected LX-2 cells to produce monocyte-attracting factors, we tested the capacity of culture supernatants from *B. abortus*-infected monocytes on MHC-I and –II expression in LX-2 cells. Culture supernatants from *B. abortus*-infected monocytes do not induce MHC-I and -II expression. However, these supernatants inhibit MHC-II expression induced by IFN-γ in an IL-10 dependent mechanism. Since hepatocytes constitute the most abundant epithelial cell in the liver, experiments were conducted to determine the contribution of these cells in antigen presentation in the context of *B. abortus* infection. Our results indicated that *B. abortus*-infected hepatocytes have an increased MHC-I expression, but MHC-II levels remain at basal levels. Overall, *B. abortus* infection induces MHC-I and -II expression in LX-2 cells, increasing the antigen presentation. Nevertheless, this response could be modulated by resident or infiltrating monocytes/macrophages.

Keywords: *Brucella*; HSC; MHC; IL-10

1. Introduction

Brucella spp. are Gram-negative intracellular bacteria that infect domestic and natural animals and produce an incapacitating chronic disease when transmitted to humans. In many countries, brucellosis remains endemic. The most frequent clinical characteristics are hepatomegaly, splenomegaly and peripheral lymphadenopathy, revealing the preference of *Brucella* for the reticuloendothelial system [1,2].

As a frequent niche of infections, the liver provides a tolerogenic environment. Such immunotolerant capacity is based on the presence of a resident immune cell repertoire in constant stimulation and the hepatic blood source that spread a unique growth factor and cytokine milieu [3].

However, the immune system of the liver is capable of inducing a prompt-response to tumor cells and pathogenic microorganisms [4]. Thus, the majority of the microorganisms that arrive in the liver are eradicated. Nonetheless, even though these several mechanisms can remove infectious agents, *Brucella* spp. can escape the immune response and persist in the liver. Accordingly, in humans infected with *Brucella*, the liver is frequently implicated, with a frequency between 5% to 52% or more [5]. Liver biopsies from *Brucella abortus*-infected patients revealed the presence of granulomas with single of multiple localizations in portal and parenchymal tissue, inflammatory infiltrations, and parenchymal necrosis [6,7].

Among the non-parenchymal cells, hepatic stellate cells (HSCs) are placed among hepatocyte and small blood vessels. They are characterized by their contents of intracellular lipid droplets and protuberances that spread nearby the blood vessels. During liver injury, HSCs are activated and realize collagen with the development of scar tissue, producing chronic fibrosis or cirrhosis [8]. Furthermore, they also have a role in liver fibrosis to heal restore inflammatory injury.

During *B. abortus* infection, the protagonism of HSCs during the generation of fibrosis has recently been revealed [9]. Besides their function during liver damage through the production of fibrosis, HSCs cans also participate as local antigen-presenting cells (APCs). HSCs express MHC class I and II molecules, as well as co-stimulatory molecules such as CD40 and CD80 [10]. Accordingly, HSCs can interact with $CD4^+$ T cells to induce effector responses [11]. In addition, HSCs direct naïve $CD4^+$ T-cell activation to T_{reg} differentiation in the presence of Dendritic cells (DC) [12]. Thus, the main role of HSCs is the ability to induce a tolerogenic liver milieu that can favor the chronicity of *B. abortus* infection.

Nucleated cells express MHC class I molecules, but MHC class II molecule expression is restricted for cell types such as dendritic cells, macrophages and B lymphocytes. MHC class II expression is regulated in part by the class II transactivator protein (CIITA) at the transcription level. The α- and β-chains of newly synthesized class II molecules are associated with the invariant change (Ii), giving rise to immature MHC-II. These molecules reach the cell surface, then recycle to the endosomal/lysosomal compartment, named MIIC. In this compartment, cathepsin S is one of the proteases responsible in Ii processing to Class II-associated invariant chain peptide (CLIP) in human antigen-presenting cells. Ii removal is an important step for the adequate export of the peptide-loaded class II molecule to the cell surface. Activation of HSCs by several agonists such as bacterial lipopolysaccharide (LPS) and IFN-γ drive the increase of MHC class II expression and co-stimulatory molecules [11]. Immune responses to liver pathogens need to consider the possibility that unconventional Antigen presenting cells (APC) play an important function, and may account for the miscarriage of effective immunity. Thus, the aim of this study is to characterize the induction of surface MHC-I and -II expression during *B. abortus* infection.

2. Results

2.1. *B. abortus Infection Induces MHC-I and -II Expression in LX-2 Cells via a T4SS-Independent Mechanism*

In this section, the capacity of *B. abortus* to induce the expression of MHC-I and -II molecules on LX-2 cells is determined. Cells were infected with *B. abortus* for 2 h, washed to eliminate the bacteria, and the infection was continued for additional 72 h. Our results indicate that *B. abortus* infection stimulated MHC-I and -II expression in LX-2 cells, yielding a level comparable to that of IFN-γ-stimulated cells used as a positive control (Figure 1).

The type 4 secretion system (T4SS) encoded by the *vir*B operon participates in the establishment of the intracellular replication niche of *Brucella* in different cell types [13], as well as contributing to the induction of a fibrotic phenotype in HSCs during *B. abortus* infection. We decided to test whether MHC-I and -II expression in infected LX-2 cells depends on a functional T4SS. No significant difference in MHC-I and -II expression was found between LX-2 cells infected with the wild-type strain and those infected with *vir*B10 isogenic mutants, indicating that the T4SS is not implicated in the modulation of MHC-I and -II expression (Figure 1). These results indicate that *B. abortus* infection induces MHC-I and -II upregulation in a T4SS-independent mechanism.

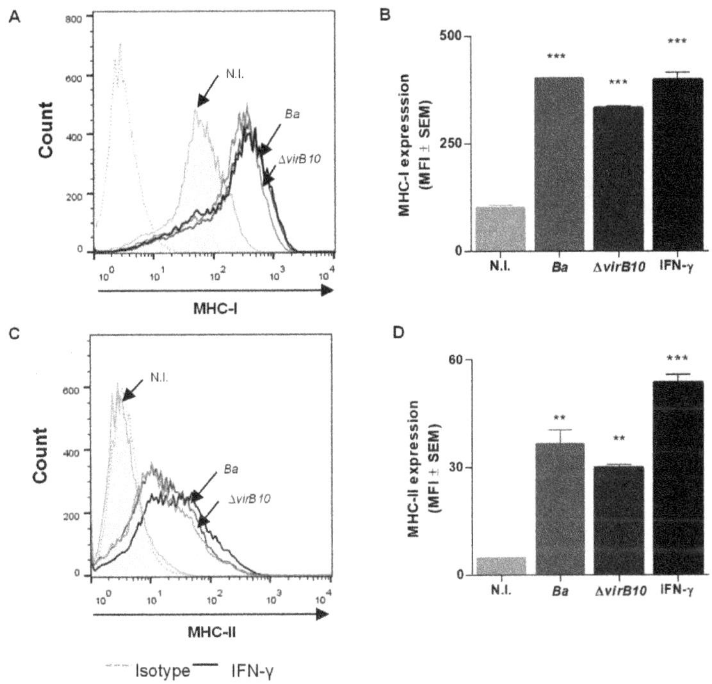

Figure 1. *Brucella abortus* infection induces MHC-I and -II expression in LX-2 cells. LX-2 cells were infected with *B. abortus* or *B. abortus virB10* mutant (Δ*virB10*) at a multiplicity of infection (MOI) of 1000 for 2 h, washed, and incubated for 72 h in complete media with antibiotics. IFN-γ (5oo U/mL) was used as a positive control. Non-infected cells (N.I.). MHC-I (**A**,**B**) and MHC-II (**C**,**D**) expression was assessed by flow cytometry. The histograms indicate the results of one representative of five independent experiments (**A**,**C**). The bars indicate the arithmetic means of five experiments, and the error bars indicate the standard errors of the means. MFI, mean fluorescence intensity (**B**,**D**). Non-specific binding was determined using a control isotype (Isotype). **, $p < 0.01$; ***, $p < 0.001$ versus non-infected cells (N.I.).

2.2. *B. abortus Infection Induce CIITA and Cathepsin S Transcription in LX-2 cells*

At transcription level, CIITA plays a key role in the MHC class II expression in professional antigen-presenting cells (APCs). Thus, experiments were conducted to evaluate whether increasing MHC-II expression correlated with increased transcription of CIITA. *B. abortus* infection induced up-regulation of CIITA mRNA after 72 h post-infection. IFN-γ was used as positive control (Figure 2A). The maturational processing of the MHC-II required the cleavage of li. The main protease involved in this process is cathepsin S [14]. Then, experiments were conducted to determine whether *B. abortus* infection was able to induce cathepsin S upregulation. *B. abortus* up-regulated cathepsin S mRNA at 72 h post-infection. IFN-γ was used as a positive control (Figure 2B). CIITA and cathepsin S primer specificity were determined by endpoint PCR (Figure 2C). These results indicate that *B. abortus* infection induces CIITA and cathepsin S transcription accordingly with the increase of MHC-II.

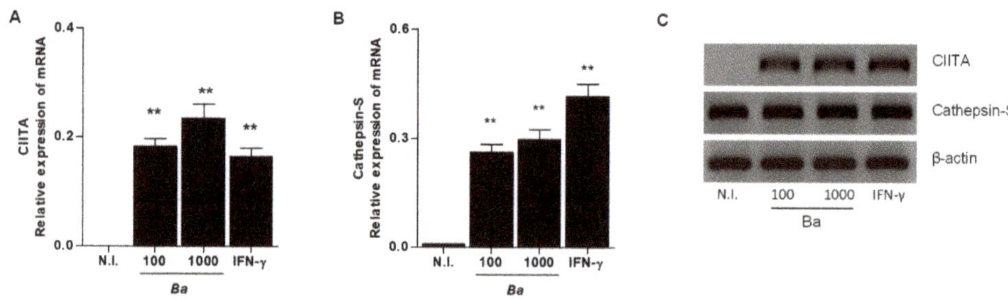

Figure 2. *B. abortus* infection induces Class II Major Histocompatibility Complex Transactivator (CIITA) and cathepsin-S expression in LX-2 cells. LX-2 cells were infected with *B. abortus* (*Ba*) at (MOIs) of 100 and 1000. At 72 h, post-infection levels of CIITA (**A**) and cathepsin-S (**B**) were determined by RT-qPCR. Agarose gel of PCR products from endpoint PCR products (**C**). Data are given as the means ± SD from three individual experiments. **, $p < 0.01$; ***, $p < 0.001$ versus non-infected cells (N.I.).

2.3. *B. abortus* Infection Does not Induce the Expression of Costimulatory Molecules CD80, CD86 and CD40

For T cells, activation the recognition of antigen/MHC complex by the T cell receptor (TCR) must be complemented by a second signal that is provided by costimulatory molecules. Experiments were conducted to determine whether *B. abortus* infection could induce CD80, CD86 and CD40 expression on LX-2 cells. *B. abortus* was unable to induce costimulatory molecule expression measured at 72 h post-infection by using specific antibodies (not shown). These results indicated that even though *B. abortus* was able to induce MHC upregulation, the costimulatory molecules remained at basal levels.

2.4. *B. abortus* Increase the Phagocytic Capability of LX-2 Cells

To determine whether *B. abortus* infection increase the phagocytic ability of LX-2 cells, cells were infected with *B. abortus* for 24 h then cultured with *Escherichia coli* for 30 min. Antibiotics were added to kill non-phagocyted *E. coli*. Counting of colony-forming unit (CFU) was performed to determine the phagocyted bacteria. *B. abortus*-infected LX-2 cells at an MOI of 1000 significantly increased phagocytosed *E. coli* in relation to uninfected cells (Figure 3A). We compared the phagocytic capacity of LX-2 cells with respect to the macrophage cell line J774.A1. Uninfected J774.A1 cells had an increased phagocytic capacity with respect to uninfected LX-2 cells. In addition, phagocytosed *E. coli* was significantly increased when J774.A1 cells were infected with *B. abortus* in an MOI-dependent fashion (Figure 3A). These results indicate that *B. abortus* infection increases the phagocytic capacity of LX-2 cells. However, its phagocytic capacity was lower than that observed in macrophages.

2.5. *B. abortus* Induces MHC-II-Restricted Antigen Processing and Presentation by LX-2 Cells

To determine if the MHC-II upregulation promoted by *B. abortus* infection was related to changes in antigen processing and the presentation of soluble antigens for MHC-II-restricted T cells, LX-2 cells were infected for 72 h then incubated with Ag85B from *Mycobacterium tuberculosis* and DB1 T-cell hybridoma, which identify soluble Ag85B processed and presented by LX-2 cells (HLA-DR1). The infection with *B. abortus* significantly increased antigen processing and presentation at multiple Ag85B concentrations, as was revealed by the increased amount of IL-2 produced by T-cell hybridoma (Figure 3B). Thus, *B. abortus* infection induces processing and presentation of soluble antigens by LX-2 cells.

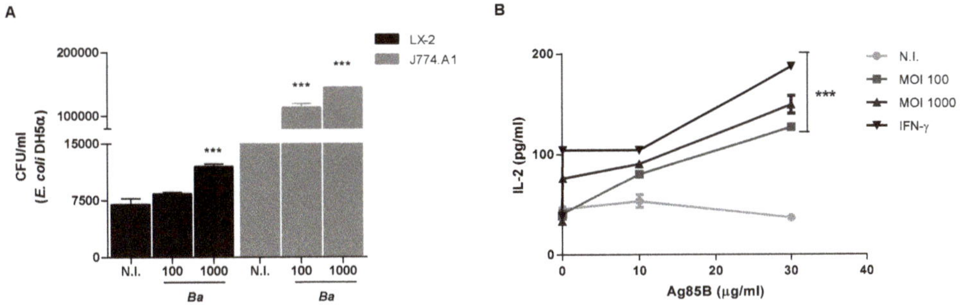

Figure 3. *B. abortus* increased the phagocytic capability and induced MHC-II-restricted processing and presentation in LX-2 cells. LX-2 cells were infected with *B. abortus* (*Ba*) at MOIs of 100 and 1000. After 72 h post-infection, *Escherichia coli* was added to the culture. Phagocytized *E. coli* were evaluated by intracellular colony-forming unit (CFU) counting (**A**); or after 72 post-infection, cells were pulsed with Ag85B for 6 h, followed by incubation with DB1 cells for 24 h. Supernatants were harvested and the amount of IL-2 was determined by ELISA (**B**). Data are given as the means ± SD from five individual experiments. ***, $p < 0.001$ versus non-infected cells (N.I.).

2.6. Culture Supernatants from B. abortus Infected THP-I Cells Do not Induce MHC-I and MHC-II Expression by LX2 Cells

In view of the capacity of *B. abortus*-infected LX-2 cells to produce chemoattractant factors of a monocyte [9] that could attract monocytes to the site of infection, we evaluated whether supernatants from *B. abortus*-infected THP-1 cells were able to modulate MHC-I and -II in LX-2 cells. To this end, LX-2 cells were treated with a 1/2 dilution of supernatants from *B. abortus*-infected and uninfected monocytes over 72 h. IFN-γ was used as a positive control. Supernatants from *B. abortus*-infected monocytes did not alter the MHC-I and -II expression levels in LX-2 cells (Figure 4A–D).

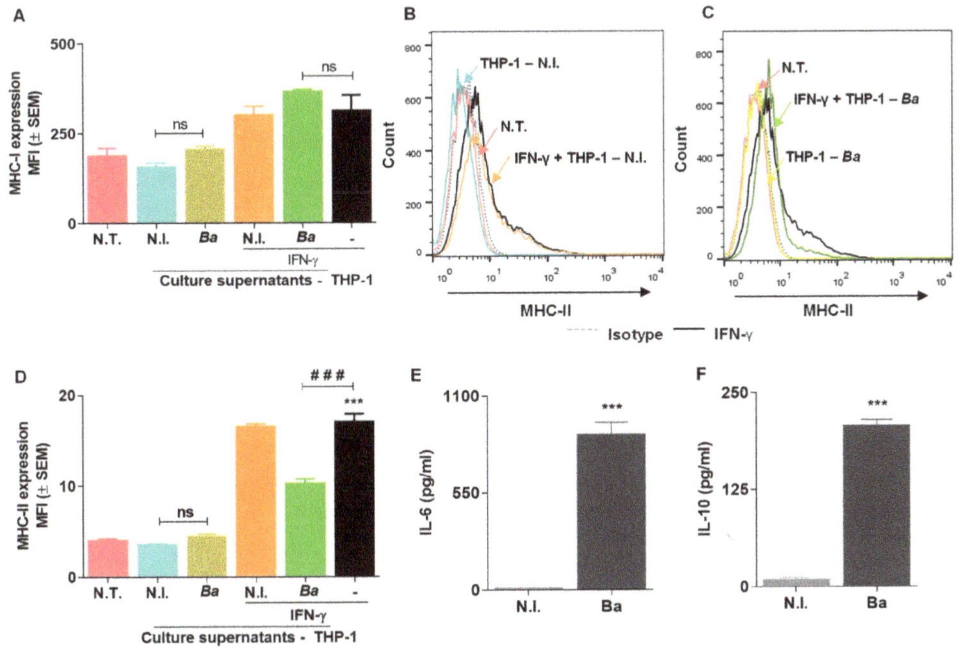

Figure 4. *Cont.*

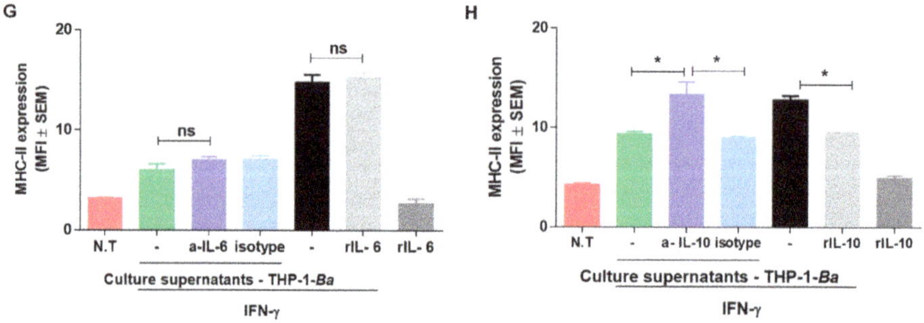

Figure 4. Modulation of MHC-I and -II expression in LX-2 cells by culture supernatants from *B. abortus*-infected monocytes. LX-2 cells were stimulated with culture supernatants from THP-1 cells infected at an MOI of 100 in the presence or not of IFN-γ (500 U/mL) or culture supernatants from uninfected THP-1 cells as control at a 1/2 dilution. After 72 h post-stimulation, MHC-I (**A**), MHC-II (**B–D**), expression was assessed by flow cytometry. IL-6 and IL-10 were determined by ELISA in culture supernatants from *B. abortus*-infected THP-1 cells at a MOI of 100 (**E**,**F**). MHC-II expression in LX-2 treated with culture supernatants from infected THP-1 cells plus IFN-γ was assayed in the presence of a neutralizing antibody anti-IL-6 (20 μg/mL), anti-IL-10 (20 ug/mL), or their isotype-matched control, with 10 ng/mL of recombinant human IL-6 (rIL-6) or 10 ng/mL of recombinant human IL-10 (rIL-10) alone or plus IFN-γ used as a control (**G**,**H**). The histograms indicate the results of one representative of five independent experiments (**A–D**,**G**,**H**). The bars indicate the arithmetic means of five experiments, and the error bars indicate the standard errors of the means. MFI, mean fluorescence intensity (**A**,**D**). Non-specific binding was determined using a control isotype (Isotype). *, $p < 0.1$; **, $p < 0.01$; ***, $p < 0.001$ versus non-infected cells (N.I.) and cells stimulated with culture supernatants from uninfected THP-1 cells.

2.7. Culture Supernatants from B. abortus Infected THP-I Monocytes Inhibit MHC-II Expression Induced by IFN-γ in an IL-10 Dependent Mechanism.

After *B. abortus* infection of macrophages, their IFN-γ-induced expression of MHC-I and -II molecules is inhibited [15,16]. Here, we have demonstrated that the *B. abortus* infection of macrophages also hits the IFN-γ-induced MHC-II but not the MHC-I expression in HSC cells (Figure 4A–D). IL-6 and IL-10 were found to be involved in the inhibition of MHC-II in different cell types, including during *B. abortus* infection [15,17], and THP-1 cells were found to secrete IL-6 and IL-10 in response to *B. abortus* infection (Figure 4E,F). Moreover, when the IL-6 and IL-10 involvement was assessed using the neutralization of specific antibodies, IL-10 but not IL-6 participated in MHC-II downregulation in LX-2 cells. In addition, IL-10 present in culture supernatants from THP-1 cells was involved in the inhibition of MHC-II expression induced by IFN-γ, since a neutralizing antibody (anti IL-10) was able to reverse the inhibitory effect (Figure 4H). In contrast, recombinant IL-6 did not inhibit MHC-II expression induced by IFN-γ in LX-2 cells (Figure 4G).

2.8. B. abortus Infection Induces MHC-I Expression in HepG2 Cells but Does not Alter MHC-II Levels

Previously we have demonstrated that *B. abortus* infects and replicates in HepG2 hepatocytes [18]. These cells represent around 60% of liver mass, and do not express MHC-II molecules under physiological conditions. However, under inflammatory conditions, hepatocytes can express MHC-II molecules and also activate T cells [19]. Experiments were conducted to determine if *B. abortus* infection is capable of modulating MHC-II expression in HepG2 hepatocytes. To this end, HepG2 cells were infected with *B. abortus* and at 72 h post-infection, and the expression of MHC-I and -II molecules were measured. *B. abortus* infection was able to differentially induce MHC-I but not MHC-II expression (Figure 5). In addition, IFN-γ was unable to induce MHC-II expression in HepG2 cells (Figure 5B).

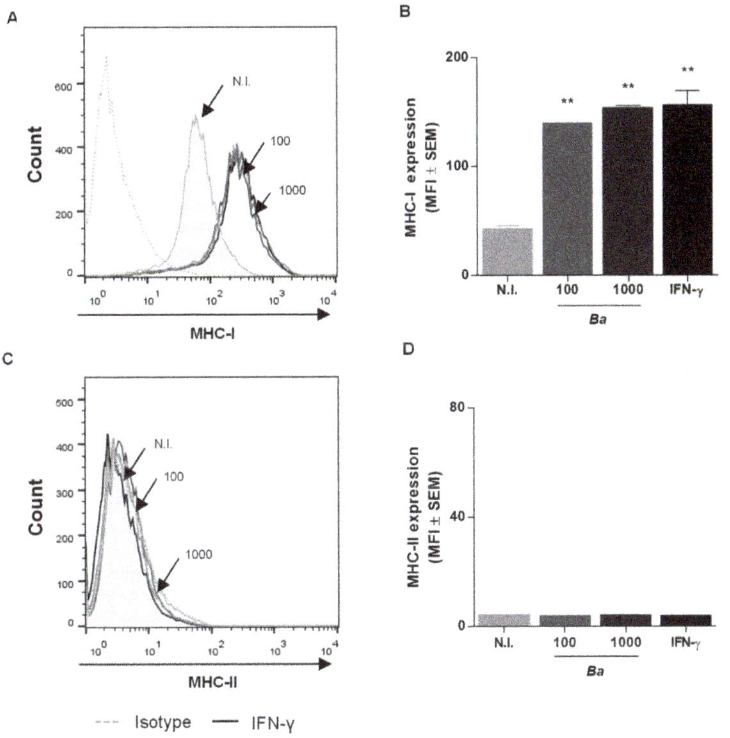

Figure 5. *B. abortus* infection induces MHC-I expression in HepG2 cells but does not alter MHC-II levels. HepG2 cells were infected with *B. abortus* at multiplicity of infections (MOIs) of 100 and 1000 for 2 h, washed, and incubated for 72 h in complete media with antibiotics. IFN-γ (500 U/mL) was used as a positive control. MHC-I (**A**,**B**) and MHC-II (**C**,**D**) expression were assessed by flow cytometry. The histograms indicate the results of one representative of five independent experiments (**A**,**C**). The bars indicate the arithmetic means of five experiments, and the error bars indicate the standard errors of the means. MFI, mean fluorescence intensity (**B**,**D**). Non-specific binding was determined using a control isotype (Isotype). **, $p < 0.01$ versus non-infected cells (N.I.).

3. Discussion

Most of the microorganisms that arrive in the liver are eliminated due to the balance between tolerance and inflammation of the hepatic microenvironment [3,20,21]. *Brucella* has a panoply of defensive strategies to evade immune response, including intracellular lifestyle and the prevention of the development of an appropriate adaptive immune response [22,23]. Thus, *Brucella* escapes from the immune response and persists in the liver, as demonstrated by the high frequency of liver pathology in human disease [5]. HSCs depict a pivotal function for wound healing of the liver [24]. Notwithstanding, the antigen-presenting capacity of HSCs has been previously reported in studies that revealed the expression of basal levels of costimulatory molecules and increases in MHC-II expression in response to IFN-γ [11,25,26]. In this study, we demonstrated the ability of *B. abortus* infection of HSCs (LX-2 cells) to upregulate MHC-I and -II expression, while the expression of the costimulatory molecules (CD80, CD86 and CD40) remained at basal levels.

The antigen presentation process involves recognition, uptake and processing by antigen-presenting cells. Previously, it has been described that the uptake of antigens by HSC is less effective than other professional APCs [27]. However, it is known that mature dendritic cells have a poor endocytic capacity, but effectively present antigens to T cells [28]. Nevertheless, *B. abortus* infection increased the efficiency of antigen uptake significantly via HSCs. Moreover, when these HSC-infected cells were cocultured with T cells, a higher level of IL-2 secretion was measured, thus inferring an increased antigen processing and further MHC-II-restricted T cell presentation after *B. abortus* infection. These results opposed other studies that have indicated that HSCs not only do not induce an effective T-cell response, but also induce the apoptosis of T cells through B7-H1 and B7-H4 signaling [26,29,30]. Such a discrepancy could be attributed to the fact that these studies eliminated the uptake, processing and presentation of antigens, since the T cell responses were performed by peptide pulsed-HSCs.

In B cells, thymus epithelial cells, and myeloid dendritic cells, CIITA is the master regulator of major histocompatibility complex (MHC) gene expression, which is constitutively expressed. However, in HSCs (among several cell types), the transcription of CIITA requires IFN-γ among others factors for both MHC-II expression [31,32] and the modulation of the transcription of MHC-I genes [33–35]. Accordingly, when HSCs are infected by *B. abortus*, the expression of MHC-I and -II are upregulated.

Antigen processing and presentation require several lysosomal proteases, including cathepsin B, L, D, and S, which are involved in the maturation of MHC-II through the processing of Ii and the cleavage of antigen peptides that will be presented [36–39]. However, the most effective proteases involved in the last step of the Ii cleavage process are cathepsin S and L. Depending on the cell type, cathepsin L and S are involved in peptide degradation [39,40]. Recently, it has been shown that cathepsin S is expressed in HSCs, which can be induced by proinflammatory cytokines such as IFN-γ. This suggests a contribution to Ii processing. In contrast, cathepsin L expression has not been significantly increased at the transcription level upon stimulation with IFN-γ [41], indicating that cathepsin S has a central role in antigen presentation in HSCs. In accordance with the increase in MHC-II expression, antigen processing, and presentation in MHC-II restricted T cells, *B. abortus* infection has also been able to induce cathepsin S mRNA transcription in HSCs.

The T4SS encoded by *virB* genes has been involved in the ability of *Brucella* to begin its intracellular replication niche [22]. In HSCs, we have previously reported that the T4SS is required to induce inflammasome activation and a fibrotic phenotype during *B. abortus* infection [42]. This system has been found to participate in the stimulation of inflammatory response during *B. abortus* infection both in vivo and in vitro [43]. However, our experiments using an isogenic a *B. abortus virB*10 polar mutant indicated that the T4SS was not involved in the induction of MHC-I and -II expression stimulated by *B. abortus* infection in HSCs.

The virulence of *B. abortus* relies on the ability of this bacteria to interact with macrophages as a central event for launching chronic *Brucella* infections [44,45]. In previous studies we have demonstrated that HSCs secrete MCP-1 in response to *B. abortus* infections [9], indicating that monocytes/macrophages could be attracted to the site of infection and, in conjunction with the resident macrophages, could modulate HSC responses. However, our results indicate that supernatants from *B. abortus*-infected macrophages were unable to induce MHC-I and -II expression.

B. abortus infection has been shown to potently activate a proinflammatory response that triggers the differentiation of T-cell responses to T-helper 1 (Th1) [46] with the simultaneous production of IFN-γ [47]. This cytokine enhances not only microbicide activities of macrophages, but also antigen-presenting functions in cells [48]. However, *B. abortus* infection can stimulate not only inflammatory but also immunomodulatory mediators such as IL-10 and IL-6 through monocytes [49,50]. These cytokines have been reported as responsible for inhibiting IFN-γ-induced MHC-II expression in immune cells [51,52]. Our experiments demonstrate that during the *B. abortus* infection of HSCs, IL-10 but not IL-6 present in supernatants from *B. abortus*-infected monocytes was implicated, at least in part, in the inhibition of IFN-γ-induced MHC-II expression.

B. abortus infection can infect and replicate in hepatocytes, inducing an inflammatory response [18]. Here, we demonstrate that in the setting of *B. abortus* infection, the MHC-I but not the MHC-II expression was induced in hepatocytes, thus enabling the hepatocytes to be susceptible to CD8+ cytotoxic T cell action.

In conclusion, the *B. abortus* infection of hepatic stellate cells and hepatocytes is able to regulate differentially the MHC expression, thus stimulating the T-cell specific-immune response at the liver. However, due to a cellular interplay, such responses may also be modified by resident or infiltrating *B. abortus*-infected monocytes/macrophages. Such bacterial skills exerted on hepatic cells may promote the evasion of immune surveillance, thus favoring its chronicity in the liver.

4. Materials and Methods

4.1. Bacterial Culture

Brucella abortus S2308 or the isogenic *B. abortus vir*B10 polar mutant (kindly provided by Diego Comerci, UNSAM University, Argentina) were cultivated in 10 ml of tryptic soy broth (Merck, Buenos Aires, Argentina) for 18 h with constant agitation at 37 °C. Bacteria were harvested and the inoculum were prepared as described previously [53]. All experiments with live *Brucella* were carried out in biosafety level 3 facilities located at the Instituto de Investigaciones Biomédicas en Retrovirus y SIDA (INBIRS).

4.2. Cell Culture

The spontaneously immortalized human hepatic stellate cell line (LX-2) was kindly provided by Dr. Scott L. Friedman (Mount Sinai School of Medicine, New York, NY, USA). LX-2 cells were maintained in Dulbecco's Modified Eagle Medium (DMEM, Life Technologies, Grand Island, NY, USA) and supplemented with 5% fetal bovine serum (FBS; Life Technologies), L-glutamine (2 mM), sodium pyruvate (1 mM), 100 U/mL penicillin, and 100 µg/mL streptomycin. The human hepatoma cell line HepG2, the murine J774.A1 cell line, and the human monocytic cell line THP-1 were obtained from the ATCC (Manassas, VA, USA) and were cultured as previously described [18]. Monocyte differentiation from THP-1 cells was achieved through cultivation in the presence of 0.05 mmol/L 1, 25-dihydroxyvitamin D3 (Calbiochem-Nova Biochem International, La Jolla, CA, USA) for 72 h. DB1 T hybridoma cells (Ag85B specific) was kindly provided by W. H. Boom (Case Western Reserve University, Cleveland, OH, USA) and was maintained in DMEM supplemented as indicated above. All cultures were grown at 37 °C and 5% CO_2.

4.3. Cellular Infection

LX-2 cells were dispensed in 24-well plates and infected with *B. abortus* S2308 or *B. abortus virB10* polar mutant at a multiplicity of infection (MOI) of 100 or 1000. HepG2 cells were infected with *B. abortus* S2308 at an MOI of 100 or 1000, and THP-1 cells at an MOI of 100. After the bacterial suspension was dispensed, the plates were centrifuged for 10 min at 2000 rpm, then incubated for 2 h at 37 °C under a 5% CO_2 atmosphere. To remove extracellular bacteria, Cells were extensively washed with DMEM then incubated in medium supplemented with 100 µg/mL gentamicin and 50 µg/mL streptomycin to kill extracellular bacteria. LX-2 cells were harvested at 72 h to determine major histocompatibility complex class I (MHC-I), MHC-II, CD40, CD80, and CD86 surface expression and CIITA and cathepsin-S gene expression. Supernatants from THP-1 cells were harvested 24 h after infection to be used as conditioned medium.

4.4. Flow Cytometry

Infected LX-2 cells, cells treated with culture supernatants at a 1/2 dilution from THP-1 cells, or recombinant human IFN-γ-treated-LX-2 cells (500 U/mL; Endogen) were washed and incubated with fluorescein isothiocyanate-labeled (FITC) anti-human HLA-DR monoclonal antibody (MAb) (clone L243; BD Bioscience, San Diego, CA, USA), FITC-labeled anti-human HLA-ABC (clone G46-2.6; BD Bioscience), phycoerythrin (PE)-labeled anti-human CD40 (clone 5C3; BD Bioscience), PE-labeled anti-human CD86 (clone 2331(FUN-1); BD Bioscience) FITC-labeled anti-human CD80 (clone 2D10; BioLegend)m or isotype-matched control antibody (Ab) for 30 min on ice. Cells were then washed, stained with 7-Amino-Actinomycin D (7-AAD; BD Biosciences) for 10 min at 4 °C in darkness, and analyzed with a FACScan flow cytometer (Becton-Dickinson, Franklin Lakes, NJ, USA), gating on viable cells (7-AAD negative cells). Data were processed using CellQuest software (Becton Dickinson). Results were expressed as mean fluorescence intensities (arithmetic means ± standard errors of the means). MHC-II expression was also assayed in the presence of a neutralizing antibody anti-IL-6 (20 µg/mL, BD Bioscience), anti-IL-10 (20 g/mL, BD Bioscience), or their isotype matched control, with 10 ng/mL of recombinant human IL-6 (rIL-6, BD Bioscience) or 10 ng/mL of recombinant human IL-10 (rIL-10, BD Bioscience) alone or plus IFN-γ used as a control.

4.5. Cytokine ELISA

The IL-2, IL-6, and IL-10 level were measured in culture supernatants by ELISA according to the manufacturer's instructions (BD Biosciences).

4.6. Phagocytosis Assays

To study the phagocytosis capability of LX-2 cells, the phagocytic uptake of *E. coli* DH5α (Invitrogen) was measured as described [54]. Briefly, cells were infected with *B. abortus* at different MOIs, as described previously. Cells were washed twice and cultured in the presence of *E. coli* for 30 min at 37 °C in 5% CO_2. Extracellular bacteria were washed and killed with gentamicin (100 mg/mL) for 30 min. Cells were washed, lysed with 0.1% (v/v) Triton X-100, plated overnight on tripteine soy broth (TSB) agar, and colony forming units (CFU) were counted. As a positive control, the same bacteria phagocytic test was assessed using the murine macrophage cell line J774.A1.

4.7. mRNA Preparation and RT-qPCR

Total cellular RNA from LX-2 cells was extracted using Quick-RNA MiniPrep Kit (Zymo Research) and 1 µg of RNA was employed to perform the reverse transcription by means of Improm-II Reverse Transcriptase (Promega). Quantitative reverse-transcription polymerase chain reaction (qRT-PCR) analysis was achieved run on a StepOne real-time PCR detection system (Life Technology) using SYBR Green as a fluorescent DNA binding dye. The conditions of the amplification reaction were the following: 10 min 95 °C, 40 cycles for 15 s at 95 °C, 58 °C for 30 s, and 72 °C for 60 s. Primer sequences used for amplification were: β-actin, forward AACAGTCCGCCTAGAAGCAC, reverse 5′-CGTTGACATCCGTAAAGACC; cathepsin-S, forward 5′-TTATGGCAGAGAAGATGTCC, reverse 5′-AAGAGGGAAAGCTAGCAATC; CIITA, forward 5′-CCGACACAGACACCATCAAC, reverse 5′-TTTTCTGCCCAACTTCTGCT. All primer sets yielded a single product of the correct size. Relative transcript levels were calculated using the ΔΔCt method using as normalizer gene β-actin.

Endpoint PCR products were subjected to electrophoresis in 1% agarose gel, stained with ethidium bromide, visualized under UV light, and photographed. In order to normalize the qRT-PCR, the β-actin gene was included as housekeeping.

4.8. Ag Processing and Presentation Assays

LX-2 cells were cultured in 96-well flat-bottom plates (10^5 cells/well) and infected with *B. abortus* or stimulated with 500 U/mL of IFN-γ (Endogen) for 72 h. Following incubation and medium remotion, the cells were widely washed prior to Ag exposure. The cells then were pulsed with Ag85B (Abcam) 1, 10, and 30 µg/mL for 6 h, followed by incubation with DB1 T hybridoma cells (10^5 cells/well). After 2 to 24 h the supernatants were harvested and the amount of interleukin-2 (IL-2) secreted by T hybridoma cells was determined by ELISA.

4.9. Statistical Analysis

One-way ANOVA, followed by a Post Hoc Tukey Test using GraphPad Prism 4.0 software, was used to perform the statistical analysis of the results. The obtained data were represented as mean ± SEM.

Author Contributions: Conceptualization, J.F.Q., and M.V.D.; methodology, P.C.A.B., A.I.P.V., and M.M.E.; investigation, P.C.A.B., A.I.P.V., and M.M.E.; writing—original draft preparation, M.V.D.; writing—review and editing, G.H.G., J.F.Q., and M.V.D.; funding acquisition, J.F.Q., and M.V.D. All authors have read and agreed to the published version of the manuscript.

Funding: This work was supported by grants from Agencia Nacional of Promoción Científica y Tecnológica (ANPCYT, Argentina), PICT 2014-1111, PICT 2015-0316, PICT 2017-2859 to MVD and PICT-2015-1921 to JFQ. Funding agencies had no role in study design, data collection and analysis, decision to publish, or preparation of the manuscript. PCAB and AIPV are recipient of a fellowship from CONICET; GHG, JFQ, and MVD are members of the Research Career of CONICET.

Acknowledgments: We thank Horacio Salomón and the staff of the Instituto de Investigaciones Biomédicas en Retrovirus y Sida (INBIRS) for their assistance with biosafety level 3 laboratory uses. We thank W. H. Boom (Case Western Reserve University, Cleveland, OH) for providing the DB1T-cell hybridoma. We also thank David H. Canaday (Case Western Reserve University) for his outstanding guidance concerning the culture of T-cell hybridomas.

Conflicts of Interest: The authors declare no conflict of interest.

References

1. Pappas, G.; Akritidis, N.; Bosilkovski, M.; Tsianos, E. Brucellosis. *N. Engl. J. Med.* **2005**, *352*, 2325–2336. [CrossRef]
2. Hayoun, M.A.; Smith, M.E.; Shorman, M. *Brucellosis*; StatPearls Publishing: Treasure Island, FL, USA, 2020.
3. Kelly, A.M.; Golden-Mason, L.; Traynor, O.; Geoghegan, J.; McEntee, G.; Hegarty, J.E.; O'Farrelly, C. Changes in hepatic immunoregulatory cytokines in patients with metastatic colorectal carcinoma: Implications for hepatic anti-tumour immunity. *Cytokine* **2006**, *35*, 171–179. [CrossRef]
4. Robinson, M.W.; Harmon, C.; O'Farrelly, C. Liver immunology and its role in inflammation and homeostasis. *Cell. Mol. Immunol.* **2016**, *13*, 267–276. [CrossRef] [PubMed]
5. Madkour, M.M. Gastrointestinal brucellosis. In *Madkour's Brucellosis*, 2nd ed.; Madkour, M.M., Ed.; Springer: Berlin/Heidelberg, Germany, 2001; pp. 150–158.
6. Akritidis, N.; Tzivras, M.; Delladetsima, I.; Stefanaki, S.; Moutsopoulos, H.M.; Pappas, G. The liver in brucellosis. *Clin. Gastroenterol. Hepatol.* **2007**, *5*, 1109–1112. [CrossRef]
7. Heller, T.; Belard, S.; Wallrauch, C.; Carretto, E.; Lissandrin, R.; Filice, C.; Brunetti, E. Patterns of hepatosplenic brucella abscesses on cross-sectional imaging: A review of clinical and imaging features. *Am. J. Trop Med. Hyg.* **2015**, *93*, 761–766. [CrossRef] [PubMed]
8. Crispe, I.N. The liver as a lymphoid organ. *Annu. Rev. Immunol.* **2009**, *27*, 147–163. [CrossRef]
9. Arriola Benitez, P.C.; Scian, R.; Comerci, D.J.; Serantes, D.R.; Vanzulli, S.; Fossati, C.A.; Giambartolomei, G.H.; Delpino, M.V. Brucella abortus induces collagen deposition and MMP-9 down-modulation in hepatic stellate cells via TGF-beta1 production. *Am. J. Pathol.* **2013**, *183*, 1918–1927. [CrossRef]
10. Thomson, A.W.; Knolle, P.A. Antigen-presenting cell function in the tolerogenic liver environment. *Nat. Rev. Immunol.* **2010**, *10*, 753–766. [CrossRef] [PubMed]
11. Winau, F.; Hegasy, G.; Weiskirchen, R.; Weber, S.; Cassan, C.; Sieling, P.A.; Modlin, R.L.; Liblau, R.S.; Gressner, A.M.; Kaufmann, S.H. Ito cells are liver-resident antigen-presenting cells for activating T cell responses. *Immunity* **2007**, *26*, 117–129. [CrossRef] [PubMed]
12. Dunham, R.M.; Thapa, M.; Velazquez, V.M.; Elrod, E.J.; Denning, T.L.; Pulendran, B.; Grakoui, A. Hepatic stellate cells preferentially induce $Foxp3^+$ regulatory T cells by production of retinoic acid. *J. Immunol.* **2013**, *190*, 2009–2016. [CrossRef]
13. Roop, R.M., II; Bellaire, B.H.; Valderas, M.W.; Cardelli, J.A. Adaptation of the Brucellae to their intracellular niche. *Mol. Microbiol.* **2004**, *52*, 621–630. [CrossRef]

14. McCormick, P.J.; Martina, J.A.; Bonifacino, J.S. Involvement of clathrin and AP-2 in the trafficking of MHC class II molecules to antigen-processing compartments. *Proc. Natl. Acad. Sci. USA* **2005**, *102*, 7910–7915. [CrossRef] [PubMed]
15. Barrionuevo, P.; Cassataro, J.; Delpino, M.V.; Zwerdling, A.; Pasquevich, K.A.; Garcia Samartino, C.; Wallach, J.C.; Fossati, C.A.; Giambartolomei, G.H. Brucella abortus inhibits major histocompatibility complex class II expression and antigen processing through interleukin-6 secretion via Toll-like receptor 2. *Infect. Immun.* **2008**, *76*, 250–262. [CrossRef]
16. Barrionuevo, P.; Delpino, M.V.; Pozner, R.G.; Velasquez, L.N.; Cassataro, J.; Giambartolomei, G.H. Brucella abortus induces intracellular retention of MHC-I molecules in human macrophages down-modulating cytotoxic CD8(+) T cell responses. *Cell. Microbiol.* **2013**, *15*, 487–502. [CrossRef] [PubMed]
17. Mittal, S.K.; Roche, P.A. Suppression of antigen presentation by IL-10. *Curr. Opin. Immunol.* **2015**, *34*, 22–27. [CrossRef] [PubMed]
18. Delpino, M.V.; Barrionuevo, P.; Scian, R.; Fossati, C.A.; Baldi, P.C. Brucella-infected hepatocytes mediate potentially tissue-damaging immune responses. *J. Hepatol.* **2010**, *53*, 145–154. [CrossRef]
19. Herkel, J.; Jagemann, B.; Wiegard, C.; Lazaro, J.F.; Lueth, S.; Kanzler, S.; Blessing, M.; Schmitt, E.; Lohse, A.W. MHC class II-expressing hepatocytes function as antigen-presenting cells and activate specific CD4 T lymphocyutes. *Hepatology* **2003**, *37*, 1079–1085. [CrossRef]
20. Golden-Mason, L.; Douek, D.C.; Koup, R.A.; Kelly, J.; Hegarty, J.E.; O'Farrelly, C. Adult human liver contains CD8pos T cells with naive phenotype, but is not a site for conventional alpha beta T cell development. *J. Immunol.* **2004**, *172*, 5980–5985. [CrossRef] [PubMed]
21. Giuffre, M.; Campigotto, M.; Campisciano, G.; Comar, M.; Croce, L.S. A story of liver and gut microbes: How does the intestinal flora affect liver disease? A review of the literature. *Am. J. Physiol. Gastrointest. Liver Physiol.* **2020**, *318*, G889–G906. [CrossRef] [PubMed]
22. Comerci, D.J.; Martinez-Lorenzo, M.J.; Sieira, R.; Gorvel, J.P.; Ugalde, R.A. Essential role of the VirB machinery in the maturation of the Brucella abortus-containing vacuole. *Cell. Microbiol.* **2001**, *3*, 159–168. [CrossRef] [PubMed]
23. De Figueiredo, P.; Ficht, T.A.; Rice-Ficht, A.; Rossetti, C.A.; Adams, L.G. Pathogenesis and immunobiology of brucellosis: Review of Brucella-host interactions. *Am. J. Pathol.* **2015**, *185*, 1505–1517. [CrossRef] [PubMed]
24. Friedman, S.L. Hepatic stellate cells: Protean, multifunctional, and enigmatic cells of the liver. *Physiol. Rev.* **2008**, *88*, 125–172. [CrossRef] [PubMed]
25. Vinas, O.; Bataller, R.; Sancho-Bru, P.; Gines, P.; Berenguer, C.; Enrich, C.; Nicolas, J.M.; Ercilla, G.; Gallart, T.; Vives, J.; et al. Human hepatic stellate cells show features of antigen-presenting cells and stimulate lymphocyte proliferation. *Hepatology* **2003**, *38*, 919–929. [CrossRef] [PubMed]
26. Yu, M.C.; Chen, C.H.; Liang, X.; Wang, L.; Gandhi, C.R.; Fung, J.J.; Lu, L.; Qian, S. Inhibition of T-cell responses by hepatic stellate cells via B7-H1-mediated T-cell apoptosis in mice. *Hepatology* **2004**, *40*, 1312–1321. [CrossRef]
27. Ebrahimkhani, M.R.; Mohar, I.; Crispe, I.N. Cross-presentation of antigen by diverse subsets of murine liver cells. *Hepatology* **2011**, *54*, 1379–1387. [CrossRef]
28. Garrett, W.S.; Chen, L.M.; Kroschewski, R.; Ebersold, M.; Turley, S.; Trombetta, S.; Galan, J.E.; Mellman, I. Developmental control of endocytosis in dendritic cells by Cdc42. *Cell* **2000**, *102*, 325–334. [CrossRef]
29. Chinnadurai, R.; Grakoui, A. B7-H4 mediates inhibition of T cell responses by activated murine hepatic stellate cells. *Hepatology* **2010**, *52*, 2177–2185. [CrossRef]
30. Charles, R.; Chou, H.S.; Wang, L.; Fung, J.J.; Lu, L.; Qian, S. Human hepatic stellate cells inhibit T-cell response through B7-H1 pathway. *Transplantation* **2013**, *96*, 17–24. [CrossRef]
31. Reith, W.; LeibundGut-Landmann, S.; Waldburger, J.M. Regulation of MHC class II gene expression by the class II transactivator. *Nat. Rev. Immunol.* **2005**, *5*, 793–806. [CrossRef]
32. Jabrane-Ferrat, N.; Nekrep, N.; Tosi, G.; Esserman, L.; Peterlin, B.M. MHC class II enhanceosome: How is the class II transactivator recruited to DNA-bound activators? *Int. Immunol.* **2003**, *15*, 467–475. [CrossRef]
33. Gobin, S.J.; Peijnenburg, A.; Keijsers, V.; van den Elsen, P.J. Site alpha is crucial for two routes of IFN gamma-induced MHC class I transactivation: The ISRE-mediated route and a novel pathway involving CIITA. *Immunity* **1997**, *6*, 601–611. [CrossRef]

34. Gobin, S.J.; Peijnenburg, A.; van Eggermond, M.; van Zutphen, M.; van den Berg, R.; van den Elsen, P.J. The RFX complex is crucial for the constitutive and CIITA-mediated transactivation of MHC class I and beta2-microglobulin genes. *Immunity* **1998**, *9*, 531–541. [CrossRef]
35. Gobin, S.J.; van Zutphen, M.; Westerheide, S.D.; Boss, J.M.; van den Elsen, P.J. The MHC-specific enhanceosome and its role in MHC class I and beta(2)-microglobulin gene transactivation. *J. Immunol.* **2001**, *167*, 5175–5184. [CrossRef] [PubMed]
36. Reyes, V.E.; Lu, S.; Humphreys, R.E. Cathepsin B cleavage of Ii from class II MHC alpha- and beta-chains. *J. Immunol.* **1991**, *146*, 3877–3880.
37. Bevec, T.; Stoka, V.; Pungercic, G.; Dolenc, I.; Turk, V. Major histocompatibility complex class II-associated p41 invariant chain fragment is a strong inhibitor of lysosomal cathepsin L. *J. Exp. Med.* **1996**, *183*, 1331–1338. [CrossRef]
38. Fineschi, B.; Sakaguchi, K.; Appella, E.; Miller, J. The proteolytic environment involved in MHC class II-restricted antigen presentation can be modulated by the p41 form of invariant chain. *J. Immunol.* **1996**, *157*, 3211–3215.
39. Riese, R.J.; Mitchell, R.N.; Villadangos, J.A.; Shi, G.P.; Palmer, J.T.; Karp, E.R.; De Sanctis, G.T.; Ploegh, H.L.; Chapman, H.A. Cathepsin S activity regulates antigen presentation and immunity. *J. Clin. Investig.* **1998**, *101*, 2351–2363. [CrossRef]
40. Nakagawa, T.; Roth, W.; Wong, P.; Nelson, A.; Farr, A.; Deussing, J.; Villadangos, J.A.; Ploegh, H.; Peters, C.; Rudensky, A.Y. Cathepsin L: Critical role in Ii degradation and CD4 T cell selection in the thymus. *Science* **1998**, *280*, 450–453. [CrossRef] [PubMed]
41. Maubach, G.; Lim, M.C.; Kumar, S.; Zhuo, L. Expression and upregulation of cathepsin S and other early molecules required for antigen presentation in activated hepatic stellate cells upon IFN-gamma treatment. *Biochim. Biophys. Acta* **2007**, *1773*, 219–231. [CrossRef] [PubMed]
42. Arriola Benitez, P.C.; Pesce Viglietti, A.I.; Gomes, M.T.R.; Oliveira, S.C.; Quarleri, J.F.; Giambartolomei, G.H.; Delpino, M.V. Brucella abortus infection elicited hepatic stellate cell-mediated fibrosis through inflammasome-dependent IL-1beta production. *Front. Immunol.* **2019**, *10*, 3036. [CrossRef] [PubMed]
43. Gomes, M.T.; Campos, P.C.; Oliveira, F.S.; Corsetti, P.P.; Bortoluci, K.R.; Cunha, L.D.; Zamboni, D.S.; Oliveira, S.C. Critical role of ASC inflammasomes and bacterial type IV secretion system in caspase-1 activation and host innate resistance to Brucella abortus infection. *J. Immunol.* **2013**, *190*, 3629–3638. [CrossRef] [PubMed]
44. Gorvel, J.P.; Moreno, E. Brucella intracellular life: From invasion to intracellular replication. *Vet. Microbiol.* **2002**, *90*, 281–297. [CrossRef]
45. Kohler, S.; Michaux-Charachon, S.; Porte, F.; Ramuz, M.; Liautard, J.P. What is the nature of the replicative niche of a stealthy bug named Brucella? *Trends Microbiol.* **2003**, *11*, 215–219. [CrossRef]
46. Zhan, Y.; Cheers, C. Endogenous gamma interferon mediates resistance to Brucella abortus infection. *Infect. Immun.* **1993**, *61*, 4899–4901. [CrossRef]
47. Dornand, J.; Gross, A.; Lafont, V.; Liautard, J.; Oliaro, J.; Liautard, J.P. The innate immune response against Brucella in humans. *Vet. Microbiol.* **2002**, *90*, 383–394. [CrossRef]
48. Schroder, K.; Hertzog, P.J.; Ravasi, T.; Hume, D.A. Interferon-gamma: An overview of signals, mechanisms and functions. *J. Leukoc. Biol.* **2004**, *75*, 163–189. [CrossRef]
49. Hop, H.T.; Reyes, A.W.B.; Huy, T.X.N.; Arayan, L.T.; Min, W.; Lee, H.J.; Rhee, M.H.; Chang, H.H.; Kim, S. Interleukin 10 suppresses lysosome-mediated killing of Brucella abortus in cultured macrophages. *J. Biol. Chem.* **2018**, *293*, 3134–3144. [CrossRef]
50. Giambartolomei, G.H.; Scian, R.; Acosta-Rodriguez, E.; Fossati, C.A.; Delpino, M.V. Brucella abortus-infected macrophages modulate T lymphocytes to promote osteoclastogenesis via IL-17. *Am. J. Pathol.* **2012**, *181*, 887–896. [CrossRef]
51. Kitamura, H.; Kamon, H.; Sawa, S.; Park, S.J.; Katunuma, N.; Ishihara, K.; Murakami, M.; Hirano, T. IL-6-STAT3 controls intracellular MHC class II alphabeta dimer level through cathepsin S activity in dendritic cells. *Immunity* **2005**, *23*, 491–502. [CrossRef]
52. Fumeaux, T.; Pugin, J. Role of interleukin-10 in the intracellular sequestration of human leukocyte antigen-DR in monocytes during septic shock. *Am. J. Respir. Crit. Care Med.* **2002**, *166*, 1475–1482. [CrossRef]

53. Scian, R.; Barrionuevo, P.; Giambartolomei, G.H.; De Simone, E.A.; Vanzulli, S.I.; Fossati, C.A.; Baldi, P.C.; Delpino, M.V. Potential role of fibroblast-like synoviocytes in joint damage induced by Brucella abortus infection through production and induction of matrix metalloproteinases. *Infect. Immun.* **2011**, *79*, 3619–3632. [CrossRef]
54. Gaikwad, S.; Agrawal-Rajput, R. Lipopolysaccharide from rhodobacter sphaeroides attenuates microglia-mediated inflammation and phagocytosis and directs regulatory T cell response. *Int. J. Inflam.* **2015**, *2015*, 361326. [CrossRef]

pathogens

MDPI

Article

Intracellular Growth and Cell Cycle Progression are Dependent on (p)ppGpp Synthetase/Hydrolase in *Brucella abortus*

Mathilde Van der Henst, Elodie Carlier and Xavier De Bolle *

Unité de Recherche en Biologie des Microorganismes (URBM), University of Namur, 61 rue de Bruxelles, 5000 Namur, Belgium; mathilde.vanderhenst@unamur.be (M.V.d.H.); elodie.carlier@unamur.be (E.C.)

* Correspondence: xavier.debolle@unamur.be; Tel.: +32-81-72-44-38

Received: 12 June 2020; Accepted: 9 July 2020; Published: 14 July 2020

Abstract: *Brucella abortus* is a pathogenic bacterium able to proliferate inside host cells. During the first steps of its trafficking, it is able to block the progression of its cell cycle, remaining at the G1 stage for several hours, before it reaches its replication niche. We hypothesized that starvation mediated by guanosine tetra- or penta-phosphate, (p)ppGpp, could be involved in the cell cycle arrest. Rsh is the (p)ppGpp synthetase/hydrolase. A *B. abortus* Δ*rsh* mutant is unable to grow in minimal medium, it is unable to survive in stationary phase in rich medium and it is unable to proliferate inside RAW 264.7 macrophages. A strain producing the heterologous constitutive (p)ppGpp hydrolase Mesh1b is also unable to proliferate inside these macrophages. Altogether, these data suggest that (p)ppGpp is necessary to allow *B. abortus* to adapt to its intracellular growth conditions. The deletion of *dksA*, proposed to mediate a part of the effect of (p)ppGpp on transcription, does not affect *B. abortus* growth in culture or inside macrophages. Expression of a gene coding for a constitutively active (p)ppGpp synthetase slows down growth in rich medium and inside macrophages. Using an mCherry–ParB fusion able to bind to the replication origin of the main chromosome of *B. abortus*, we observed that expression of the constitutive (p)ppGpp synthetase gene generates an accumulation of bacteria at the G1 phase. We thus propose that (p)ppGpp accumulation could be one of the factors contributing to the G1 arrest observed for *B. abortus* in RAW 264.7 macrophages.

Keywords: Brucella; cell cycle; (p)ppGpp; *rsh*

1. Introduction

Bacteria from the *Brucella* genus are the causative agents of brucellosis, a neglected disease which constitutes a worldwide anthropozoonosis. *Brucella* spp. are Gram negative alphaproteobacteria belonging to the Rhizobiales order [1]. *Brucella abortus* causes severe symptoms in mammals such as abortion in pregnant females and sterility in males. In humans, the disease is characterized by an undulant fever, also named Malta fever, and in the long term the infection leads to chronicity and symptoms such as arthritis, endocarditis and can have a fatal outcome without treatment [1]. In their hosts, *Brucellae* invade, survive and replicate inside professional and non-professional phagocytic cells such as macrophages and trophoblasts. Inside host cells, *Brucellae* are found in vacuoles named *Brucella* containing vacuoles (BCVs). In the first part of their trafficking, they successively harbor markers of early and late endosomes, a phase of the trafficking in which the bacterium does not proliferate [2,3]. This compartment presents a pH of about 4.0 to 4.5 and this acidification is essential for the successful establishment of *B. suis* infection [4]. Afterwards, the bacteria are found in BCVs having markers of the endoplasmic reticulum, where they replicate [5]. Later in the cellular infection, bacteria are found in vacuoles characterized by autophagy-related proteins [6].

Different cellular models for in vitro study of *B. abortus* infection of have been developed, such as the use of RAW 264.7 macrophages and HeLa epithelial cells. Some years ago, the investigations of the *B. abortus* infection process in these models revealed that the cell cycle regulation of *B. abortus* is linked to its virulence [2]. Indeed, for bacteria that did not segregate duplicated replication origins, the so-called G1 cells are more infectious than the S or G2 phase bacteria, i.e., bacteria currently replicating their genome or at the stage between the completion of genome replication genome and cell vision, respectively. More importantly, after internalization, bacteria remain in the G1 stage for up to 8 h, depending the host cell type, in Lamp1-positive compartments before reaching the endoplasmic reticulum where *B. abortus* can restart its cell cycle, its DNA replication and actively proliferate [2]. During the first hours of the infection, in BCVs with endocytic markers, *B. abortus* encounters harsh conditions such as acidic stress [7] and alkylating stress [8]. In addition, it was already proposed that *B. abortus* has to face a starvation conditions inside host cells [9]. Starvation is the most obvious condition that could explain why *B. abortus* is blocked at the G1 stage of the cell cycle during the first phase of its intracellular trafficking in HeLa cells and RAW 264.7 macrophages. Starvation sensing is classically involving the synthesis of (p)ppGpp (guanosine penta- or tetra-phosphate), also called alarmone. The synthesis and degradation of (p)ppGpp are catalyzed by enzymes of the RelA/SpoT family, also called Rsh enzymes. It was found that *rsh* mutants, which should be not able to produce (p)ppGpp anymore, are strongly impaired during in vitro infection as well as during murine infection [10,11].

The alarmone (p)ppGpp is widely used by bacteria to quickly adapt to stress conditions such as nutrient starvation. The production and accumulation of this alarmone induces pleiotropic effects, modulating transcription and translation, that commonly result in cell cycle and DNA replication delay [12–15]. The ability to produce (p)ppGpp has been associated with virulence in bacterial pathogens belonging to relatively distant phylogenetic groups, such as *Legionella pneumophila* [16], *Vibrio cholerae* [17], and *Mycobacterium tuberculosis* [18]. In *Escherichia coli*, during the stringent response induced by starvation, (p)ppGpp binds directly to a site located at the interface between the β' and ω subunits of the RNA polymerase [19]. A second distinct site between the β' subunit and the DksA transcription factor has been shown to be bound by (p)ppGpp as well [20]. This interaction has been shown to enhance the transcriptional effects of DksA on the RNA polymerase, suggesting synergistic effects of DksA and (p)ppGpp together [20].

The RelA/SpoT homolog proteins are responsible for (p)ppGpp homeostasis. In *E. coli*, there are two enzymes of the Rsh family, RelA and SpoT [21]. SpoT contains a synthetase domain, a hydrolase domain, and two C-terminal regulatory domains; thus, this enzyme can both catalyze the production and the degradation of (p)ppGpp, respectively. RelA contains similar domains, however the functionality of the hydrolase domain of RelA has been lost during evolution, leading to a monofunctional enzyme that can only synthesize the alarmone [22]. In most alphaproteobacteria, including *B. abortus*, the production and the degradation of (p)ppGpp depends on one enzyme named Rsh (for RelA SpoT homolog) [10,21].

In the present study, we analyzed the impact of alterations in (p)ppGpp synthesis or degradation on the growth, the cell cycle and the infection process of *B. abortus*. We show that mutants either unable to produce (p)ppGpp or producing a (p)ppGpp hydrolase are impaired for the infection process. In addition, our results show that expression of a constitutive (p)ppGpp synthetase negatively impacts growth and DNA replication of *B. abortus*, and also leads to a strong proliferation defect during infection of RAW 264.7 macrophages. We also observed that a *B. abortus dksA* null mutant was able to proliferate inside host cells as the wild type (WT) strain, suggesting that DksA is not crucially involved in the (p)ppGpp-dependent phenotypes observed during infection. These results suggest that adjustment of (p)ppGpp levels are crucial for the infection process in *B. abortus*.

2. Results

2.1. *rsh* Deletion Drastically Impacts Growth in Minimal Medium and the Infection Process

We generated a Δ*rsh* strain by allelic replacement in *B. abortus* 544 and we assayed the growth of this strain in rich culture medium (2YT) as well as in Plommet minimal medium [23] supplemented with erythritol as a carbon source. The growth of Δ*rsh* in 2YT was similar to the WT strain during the exponential phase, but the shift into the stationary phase occurred later and at a higher optical density (OD) compared to the wild type strain (Figure 1A). To evaluate bacterial viability, we counted the colony forming units (CFUs) throughout the culture in liquid rich medium (Figure 2). The Δ*rsh* strain showed a clear survival defect during the stationary phase, marked by a decrease in CFUs between 24 h and 48 h compared to the WT and the complemented strain.

Because it has been shown in other bacteria that the stringent response is linked to nutrient availability, we tested the growth of the Δ*rsh* strain in Plommet minimal medium, supplemented with erythritol as a carbon source, to mimic starvation conditions. The Δ*rsh* showed a clear growth defect compared to the WT as the OD rapidly decreased during the mid-exponential phase, indicating that Δ*rsh* cannot grow and survive in this medium, as expected for a mutant unable to produce (p)ppGpp (Figure 1B).

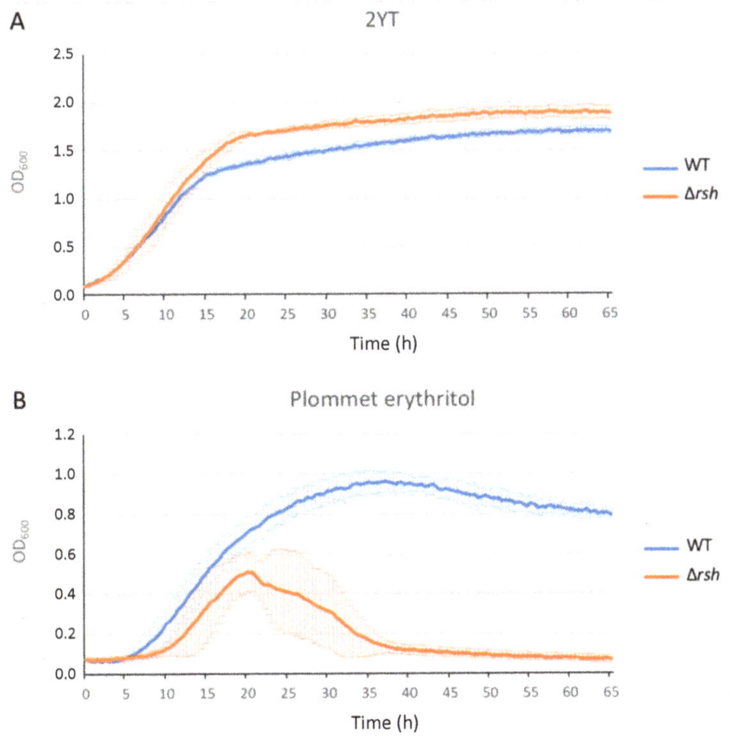

Figure 1. Growth of the Δ*rsh* mutant in 2YT rich medium (**A**) and in Plommet erythritol minimal medium (**B**). Strains were grown in liquid culture overnight in order to reach exponential phase. Cultures were then diluted at an optical density (OD) of 0.1 in 2YT medium. The OD of each strain was measured every 30 min. The graph represents the means of a biological quadruplicate. The error bars represent the standard deviation for each time point. WT: wild type.

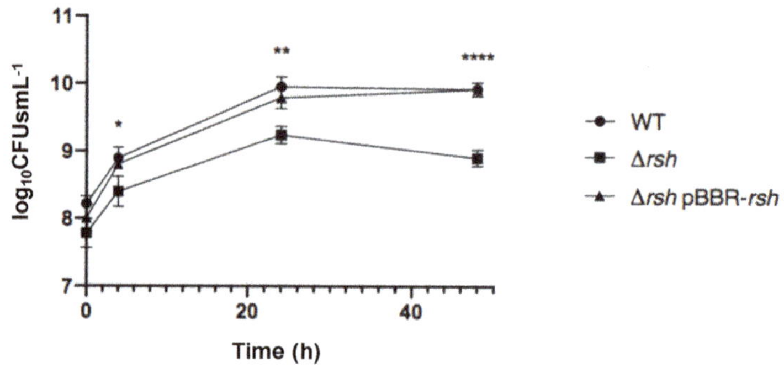

Figure 2. Survival and growth of *B. abortus* WT, Δ*rsh* and Δ*rsh* pBBR-*rsh* in 2YT rich medium. Strains were grown in liquid culture overnight in order to reach exponential phase. Cultures were then diluted at an OD of 0.1 (3×10^8 bacteria/mL for the WT strain) in 2YT medium. The numbers of live bacteria (log_{10}CFUs mL^{-1}) were determined at 0 h, 4 h, 24 h and 48 h by plating serial dilutions. Values represent the means of three independent experiments and the error bars represent the standard deviation. The asterisks mean significant for $p < 0.05$ (*) $p < 0.01$ (**); $p < 0.0001$ (****), and the p values were calculated by one-way ANOVA.

We tested the ability of Δ*rsh* to infect and multiply inside RAW 264.7 macrophages compared to the WT strain and the complemented strain by performing CFU counting throughout the cellular infection (Figure 3). The Δ*rsh* mutant showed a significant decrease in CFUs at 24 h post-infection compared to the WT strain, suggesting that the *rsh* gene is required for intracellular proliferation. A slight but significant difference was also observed at 2 h post-infection between the WT and complemented strain (Figure 3), probably highlighting a low toxicity of the vector or a *rsh* overexpression effect.

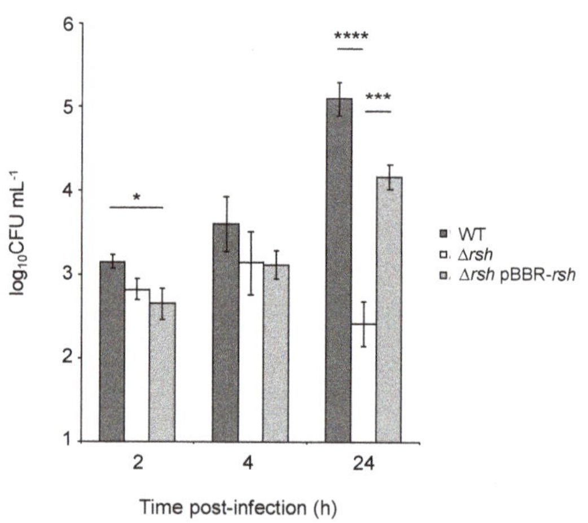

Figure 3. Survival and growth of *B. abortus* WT, Δ*rsh* and Δ*rsh* pBBR-*rsh* during infection of RAW 264.7 macrophages. Strains were grown in liquid culture overnight in order to reach exponential phase. Cultures were then diluted in Dulbecco's Modified Eagle's Medium (DMEM) to obtain a multiplicity of infection (MOI) of 50. The numbers of live bacteria (log_{10}CFUs mL^{-1} of cellular lysate, 0.5 mL per well) were determined at 2 h, 4 h, and 24 h by plating serial dilutions. Values represent the means of three independent experiments and the error bars represent the standard deviations. A one-way ANOVA test was performed as statistical analysis. The asterisks mean significant for $p < 0.05$ (*); $p < 0.001$ (***); $p < 0.0001$ (****).

2.2. The Artificial Hydrolysis of (p)ppGpp Leads to a Δrsh Phenotype during Infection

Since Rsh is responsible for (p)ppGpp homeostasis and an Δ*rsh* mutant failed to proliferate inside RAW 264.7 cells, we tested the involvement of (p)ppGpp in the infection process. However, it is known that Rsh is involved in regulation networks through protein–protein contacts [14,24] in other bacteria. Therefore, we cannot rule out that the absence of the Rsh protein, rather than the absence of (p)ppGpp, would be responsible for the defect observed in infection. This is reinforced by the observation that mutants for homologs of the glutamine-dependent control pathway of Rsh are also attenuated in RAW 264.7 macrophages [25]. We thus generated a strain in which (p)ppGpp is hydrolyzed by a strong (p)ppGpp hydrolase, a product of the *mesh1* gene from *Drosophila melanogaster* [26]. Indeed, it was shown that Mesh1 was active in vitro and in vivo [26]. We thus expect this heterologous enzyme to be constitutive in *B. abortus*. We adapted the *mesh1* coding sequence for the codon bias of *B. abortus* and expressed the resulting coding sequence on a medium copy replicative plasmid, leading to the *B. abortus* pBBRi-*mesh1b* strain. Interestingly, this strain showed a clear decrease in CFUs at 24 h post-infection of RAW 264.7 macrophages (Figure 4), which is consistent with a crucial role played by (p)ppGpp to allow growth inside host cells, as suggested above.

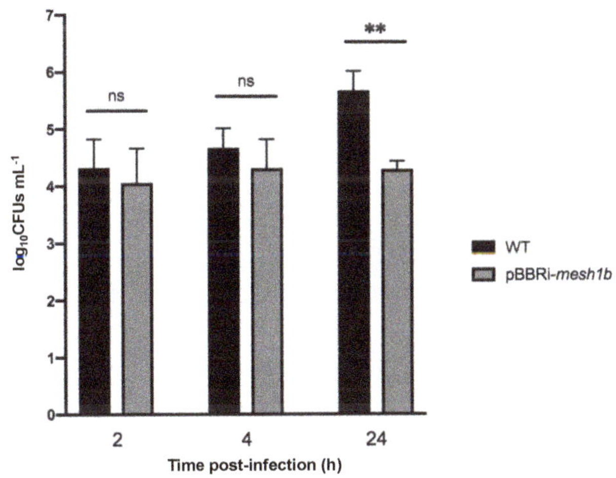

Figure 4. Survival and growth of *B. abortus* WT and pBBRi-*mesh1b* during infection of RAW 264.7 macrophages. Strains were grown in liquid culture overnight in order to reach exponential phase. Cultures were then diluted in DMEM to obtain a MOI of 50. The numbers of live bacteria (log_{10}CFUs mL^{-1}) were determined at 2 h, 4 h, and 24 h post-infection by plating serial dilutions. Values represent the means of three independent experiments and the error bars represent the standard deviation. A Student *t* test was performed for the comparison of the two strains. The asterisks mean significant for $p < 0.01$ (**) and "ns" means "not significant".

2.3. Expression of a Constitutive Allele for a (p)ppGpp Synthetase Impacts Bacterial Growth and Chromosome Replication

In order to get more insight about the role of (p)ppGpp in *B. abortus*, we constructed a strain that artificially produces this alarmone. We used a truncated version of the *relA* gene from *E. coli*, *relA'* [12] that removes the C-terminal regulatory domains of the encoded protein. The *relA'* coding sequence was inserted downstream of an isopropyl β-D-1-thiogalactoside (IPTG)-inducible promoter on the pSRK replicative plasmid [27]. The resulting strain, named *pSRK-relA'*, is supposed to produce (p)ppGpp synthetase when IPTG is added to the medium. As a negative control, we used the *pSRK-relA'** strain containing the point mutation E335Q, which leads to a catalytically dead protein. Since the detection of (p)ppGpp levels using ^{32}P is not compatible with our biosafety level 3 set up, we tried to gain indirect evidence that (p)ppGpp is indeed produced when the expression of *relA′* is induced. We assayed the growth of the *pSRK-relA′* and *pSRK-relA′** strains in rich culture medium with or without IPTG induction. The *pSRK-relA′*, *pSRK-relA′** and WT strains grew equally in 2YT; however, when IPTG was added to the medium, a growth delay was only observed for the *pSRK-relA′* strain (Figure 5). This observation is consistent with the production of (p)ppGpp levels that are sufficient to limit growth when *relA′* expression is induced.

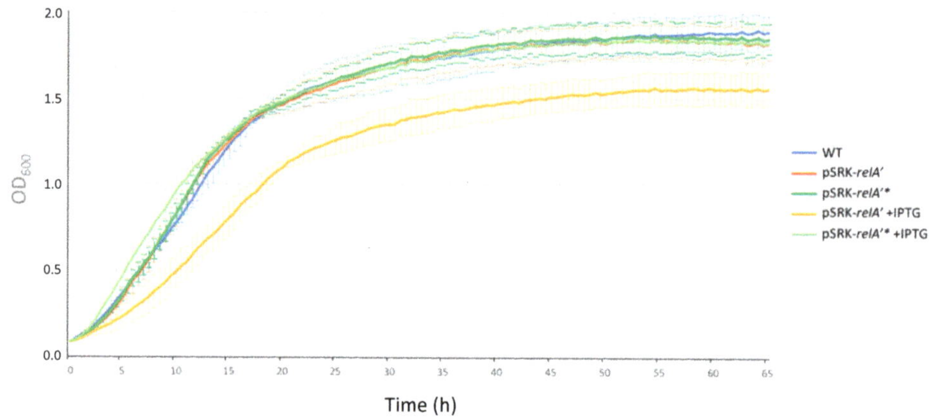

Figure 5. Growth curve in rich culture medium for the WT, *pSRK-relA′* and *pSRK-relA′** with or without IPTG. Strains were grown in liquid culture (2YT medium) overnight in order to reach exponential phase. Cultures were then diluted at an OD of 0.1 in 2YT medium supplemented or not with IPTG. The OD of each strain was measured every 30 min. The graph represents the means of a biological triplicate. The error bars represent the standard deviation for 3 biological replicates for each time point.

Since it was already reported that (p)ppGpp has an impact on DNA replication in *C. crescentus* and *E. coli* [12–15], we took advantage of a *B. abortus* strain allowing us to monitor the chromosomal replication status at the single cell level in order to study the impact of (p)ppGpp overproduction on DNA replication. This strain expresses an *mCherry-parB* allele that allows us to highlight the segregation of replication origin(s) of chromosome I. In this strain, one mCherry focus means that segregation has not yet started and the bacterium is probably in the G1 phase of the cell cycle, and two mCherry foci correspond to two segregated replication origins, meaning that the bacterium has already started replication and is thus in the S or G2 phase of the cell cycle [2]. The *pSRK-relA′* and *pSRK-relA′** plasmids were inserted in a *B. abortus mCherry-parB* strain and we counted the number of G1 bacteria every two hours for 6 h after the inoculation of bacteria in rich medium, with or without IPTG. Interestingly, we observed an increase in the proportion of G1 bacteria over the time of induction with IPTG for the *pSRK-relA′* strain (Supplementary Figure S3). The proportion of G1 bacteria of the non-induced *pSRK-relA′* and both the induced or non-induced *pSRK-relA′** remained stable after the addition of IPTG (Figure 6). These results strongly suggested that artificial induction of (p)ppGpp synthesis could delay the transition between the G1 phase to the S phase and subsequently have an impact on the initiation of chromosomal replication in *B. abortus*.

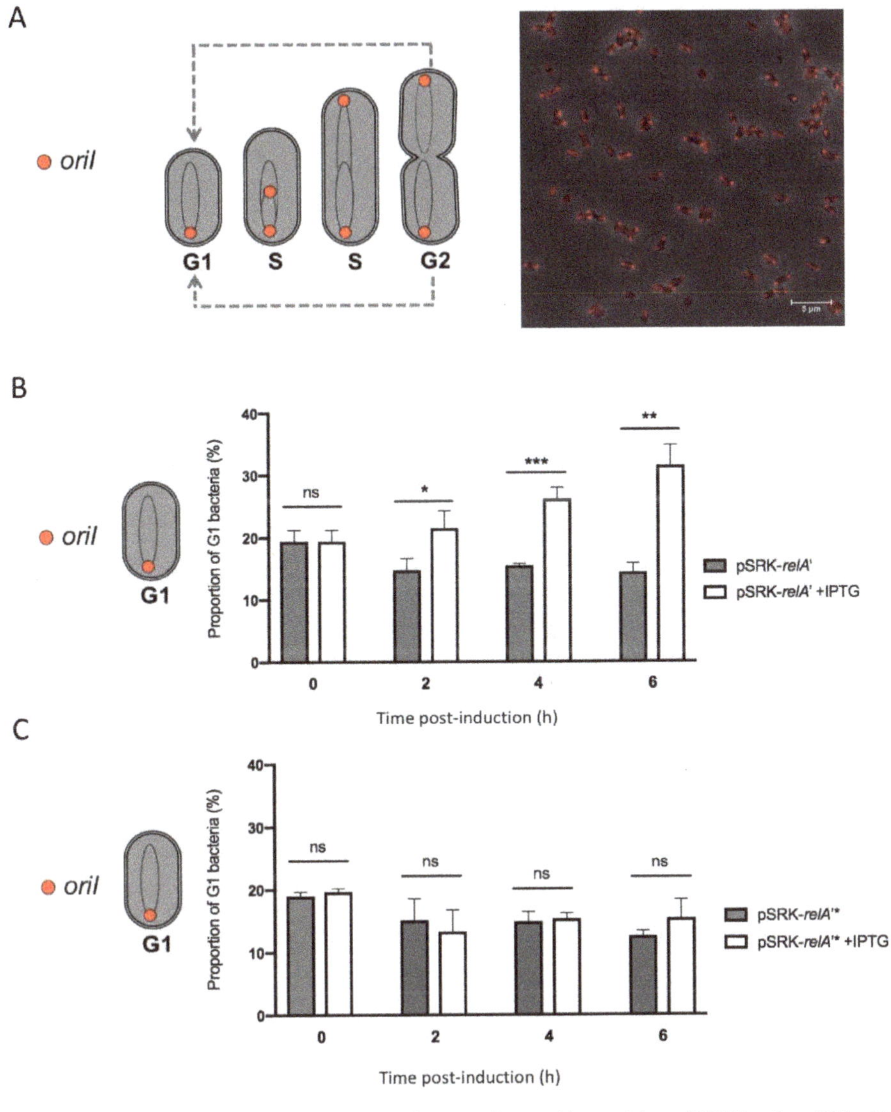

Figure 6. Proportion of G1 bacteria in rich culture medium with or without IPTG for the *pSRK-relA'* and *pSRK-relA'** strains. (**A**) Schematic drawing of the mCherry-ParB localization throughout the cell cycle [2] and fluorescence microscopy of the *pSRK-relA' mCherry-parB* strain. Scale bar represents 5 µm. (**B**) Strains were grown in liquid culture (2YT medium) overnight in order to reach exponential phase. Cultures were then diluted to an OD of 0.1 in 2YT medium supplemented or not with IPTG. Samples were taken every 2 h, placed on a phosphate-buffered saline (PBS) agarose pad and observed with a fluorescence microscope. Bacteria in G1 phase (presenting only one focus of mCherry-ParB) were counted for each time post-induction. Error bars represent the standard deviation from the means of three independent experiments (biological triplicates). The significant differences are indicated by $p < 0.05$ (*), $p < 0.01$ (**) and $p < 0.001$ (***); "ns" means not significant. The number of bacteria considered in these triplicate experiments are detailed in Supplementary Table S1.

2.4. *Induced Production of a Constitutive (p)ppGpp Synthetase Leads to a Proliferation Defect during Infection*

Since (p)ppGpp overproduction seemed to have an impact on replication, i.e., an increase of the proportion of G1 cells in the bacterial population, and that the G1 bacteria are more infectious, we decided to investigate the effect of overproduction of (p)ppGpp on the infection process. We infected RAW 264.7 macrophages with the *pSRK-relA'* strain induced or not with IPTG (Figure 7). The IPTG was kept in the cell culture medium during the infection for the induced condition. We first observed that bacterial internalization is not enhanced by the increase in the proportion of bacteria in the G1 phase of the cell cycle. We also observed that induction of *pSRK-relA'* induced a strong defect in intracellular proliferation compared to the WT and uninduced *pSRK-relA'* conditions. This result suggests that overproduction of (p)ppGpp during infection prevents growth in the intracellular niche.

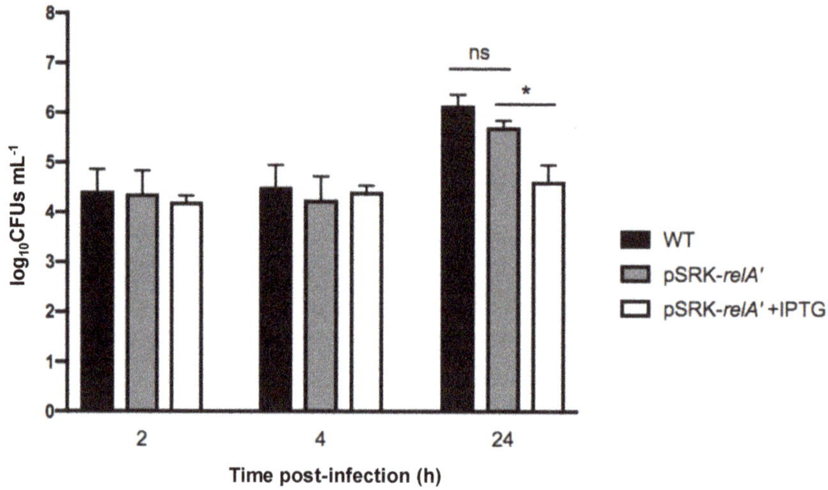

Figure 7. Survival of *B. abortus* WT and *pSRK-relA'* strains with and without IPTG during infection of RAW 264.7 macrophages. Strains were grown in liquid culture overnight in order to reach exponential phase. Cultures were then diluted at an OD of 0.1 with or without IPTG (1 mM) and were incubated for 3 h at 37 °C. Cultures were then diluted in DMEM with or without IPTG (10 mM) to obtain a MOI of 50. Concentrations of live bacteria ($\log_{10}$CFUs mL^{-1}) were determined at 0 h, 4 h, and 24 h post-infection by plating serial dilutions. Values represent the means of three independent experiments and the error bars represent the standard deviation. A Student's t test was performed as statistical analysis. The asterisks mean significant for $p < 0.05$ (*) and "ns" means "not significant".

2.5. *DksA Is Not Required during the Infection Process*

Because (p)ppGpp seemed important during host infection and DksA is involved in a part of the (p)ppGpp transcriptional response in other species, we tested the ability of Δ*dksA* to infect and proliferate inside RAW 264.7 macrophages. No difference in CFUs was observed between WT and Δ*dksA* strains (Supplementary Figure S1), meaning that DksA is not crucially involved in the infection process and that the phenotype observed for (p)ppGpp-deprived mutants (Δ*rsh* and *pBBRi-mesh1b*) is probably not mediated by DksA.

3. Discussion

Brucella abortus is able to control its cell cycle progression when it is inside host cells, particularly the replication and segregation of its replication origins [2]. However, the molecular mechanisms involved in this control are unknown. Since convincing data show that the ability to adapt to starvation is a key factor for the success of cellular infections by *Brucella melitensis* and *Brucella suis* [10,11], we further investigate the role of the (p)ppGpp, the alarmone produced in the presence of starvation conditions, and the Rsh enzyme that is proposed to synthesize this alarmone. We first confirmed that in *B. abortus* 544, like in other *Brucella* strains, Rsh is crucial for the success of a cellular infection (Figure 3), for the survival in stationary phase (Figure 2) and the growth in minimal medium (Figure 1). In agreement with the absence of proliferation of the *rsh* mutant in macrophages, a strain constitutively producing a (p)ppGpp hydrolase (Mesh1b) from *Drosophila melanogaster* is also unable to grow in RAW 264.7 macrophages. We observed that the survival of *pBBR-mesh1b* strain is less severely impacted during infection than the Δ*rsh* strain. One could imagine that this intermediate phenotype is due to the presence of residual alarmone in the *pBBR-mesh1b* strain, while it is not the case in the Δ*rsh* strain since the (p)ppGpp synthetase domain is not present. Another explanation for these different phenotypes could be that Rsh plays additional role(s) for survival in infection than the regulation of (p)ppGpp homeostasis. Interestingly, a mutant overproducing (p)ppGpp is also unable to proliferate in these cells, suggesting that the (p)ppGpp level should be in a specific range of concentration to allow cellular infection; having too much or not enough (p)ppGpp would be detrimental for the success of the cellular infection.

What does (p)ppGpp control and how is Rsh regulated? It was shown that the absence of *rsh* in *B. suis* affects the transcription of genes known to be involved in virulence [28], such as *pyrB*, which was shown to be essential for *B. abortus* proliferation in RAW 264.7 macrophages [25]. It is thus likely that a minimum level of (p)ppGpp would be required for the success of a cellular infection by *B. abortus*. It was shown that the glutamine pool modulates Rsh (called SpoT) in the model alpha-proteobacterium *Caulobacter crescentus* through the phosphotransferase system (PTS) [14]. Interestingly, mutants for components of this system were found to be attenuated in RAW 264.7 macrophages [25]. Moreover, PTS and the two-component regulator BvrR control the expression of the *virB* operon [29,30], coding for a type IV secretion system that is crucial for intracellular proliferation in most cell types [31]. These data indicate that a quite complex regulation network is probably linking Rsh control and virulence. However, the molecular mechanisms controlling Rsh activity in *B. abortus* are unknown and deserve further investigation.

One striking conclusion of our data is the moderate effect that (p)ppGpp overproduction has on the proportion of bacteria at the G1 stage of the cell cycle. Indeed, while overproduction seems to be sufficient to impair growth inside host cells (Figure 7), the proportion of G1 after 6 h of induction is about a third of the culture while it is approximately 15% in the absence of induction (Figure 6). Since the cell cycle takes about 3 h in the conditions tested, it is likely that only a fraction of the bacteria arrested their cell cycle at the G1 stage. Inside host cells, the proportion of G1 cells is about 75% and remains stable for 2 to 4 h at least [2], suggesting that other mechanisms are probably involved in the control of the cell cycle in host cells, early in the trafficking. These mechanisms could involve the acidic nature of the BCV, or the diffusion sensing proposed to occur through a regulation system homologous to quorum sensing [32]. More investigations are thus needed to discover the multiple factors involved in the cell cycle control of *B. abortus* inside host cells.

4. Materials and Methods

4.1. Strains and Growth Conditions

The reference strain *B. abortus* 544 was used for all experiments and was grown on solid or in liquid 2YT medium (LB 32 g/L Invitrogen, Yeast Extract 5g/L, BD and Peptone 6 g/L, BD) at 37 °C. *E. coli* strain DH10B was used for plasmid constructs and the conjugative strain *E. coli* S17-1 was used for mating with *B. abortus*. Both strains were cultivated in LB medium (Luria Bertani, Casein Hydrolysate 10g/L, NaCl 5g/L, Yeast Extract 5g/L) at 37 °C. Depending on the plasmid used, different selection antibiotics were added to the culture medium: ampicillin (100 μg/mL); carbenicillin (100 μg/mL); kanamycin (50 μg/mL for the replicative plasmid, and 10 μg/mL for the integrated plasmid); nalidixic acid (25 μL/mL); chloramphenicol (20 μg/mL for the replicative plasmid and 4 μg/mL for the integrated plasmid). Isopropyl β-d-1-thiogalactopyranoside (IPTG) was used at a concentration of 1 mM in bacterial liquid culture and at 10 mM in the mammalian cell culture medium during cellular infections. When the Δ*rsh* mutant was constructed, we added casamino acids 0.5% (BactoTM Casamino Acids from Thermo Fisher, Waltham, MA, USA) in the conjugation medium.

4.2. Strains Construction

Deletion strains were constructed by allelic exchange using the pNPTS138 vectors (M. R. K. Alley, Imperial College of Science, London, UK) carrying the upstream and the downstream regions of the targeted gene. The primer sequences used for amplification of the upstream region of the *dksA* gene were 5′-ttGGATCCcaagcgccagatcttca-3′ and 5′-ttGAATTCttcactcattctgaatcacccc-3′. The primer sequences used for amplification of the downstream region of the *dksA* gene were 5′-ttGAATTCtgatatcgaataatggtttggaaa-3′ and 5′-ttAAGCTTcgcccagcttcaaattac-3′.

We used the *rsh* deletion plasmid pMQ203 (provided by M. Quebatte, Biozentrum, Basel), containing the upstream and downstream regions of *rsh* amplified with the following hybridization sequences: 5′-ccggatgatctgaaggaa-3′, 5′-gcgcatcatctgccgaaa-3′ and 5′-gtctgggacctcaagcat-3′, 5′-cccgtggtgacgatatct-3′.

The Δ*rsh* pBBR-*rsh* strain was generated by inserting the pBBR-*rsh* in the Δ*rsh* strain. The pBBR-*rsh* was constructed by cloning the endogenous promoter of *rsh* and the *rsh* coding sequence, amplified with the primers 5′-aaaCTCGAGgcgagattgccgatgaga-3′ and 5′-aaaCTGCAGctatccgttcacacgctttg-3′.

The pBBRi-*mesh1b* strain was constructed by inserting the coding sequence *mesh1b* in the pBBRi plasmid. The sequence of *mesh1b* was adapted to the codon usage of *Brucella*, and is available in Supplementary Figure S2.

The *mCherry-parB* strains containing pSRK-*relA′* and pSRK-*relA′** were created using the Tn7 system [33] which consists in transposition of mini-Tn7 expressing *mCherry-parB* under the control of the *PgidA* promoter as previously reported [2] and the resistance cassette to ampicillin/carbenicillin under the control of P*bla* promoter at the *glmS* locus of *B. abortus*. The primer sequences used for the amplification of P*gidA*-mCherry-*parB* and P*bla-amp* were 5′-cgcggatcctctgtggaatcctgtttgttg-3′, 5′-AGCGGATACATATTTGAActagctttgaagacggcg-3′ and 5′-TTCAAATATGTATCCGCTCATGA-3′, 5′-cgggatccTTACCAATGCTTAATCAGTGAGG-3′.

4.3. Growth Assays

The bacterial growth curves were performed using a bioscreen (Epoch2 Microplate *Photospectrometer* from BioTek). Bacterial cultures in the exponential phase of growth were washed two times with PBS and were normalized at an OD of 0.1 in a given medium. A 200 μL aliquot of the normalized culture was transferred to a plate and each condition was performed in technical triplicate (3× 200 μL). The plate was incubated at 37 °C with shaking and the OD_{600} of each well was measured every 30 min. One biological replicate constitutes the mean of three technical replicates and experiments were repeated at least three times to obtain biological triplicates.

4.4. Survival Assays

Bacterial cultures in the exponential phase of growth were normalized at an OD of 0.1 in 2YT liquid medium. Serial dilutions were plated on 2YT solid medium at different time points and plates were incubated at 37 °C.

4.5. Infections of RAW 264.7 Macrophages

RAW macrophages were put in wells in DMEM medium (with decomplemented bovine serum, glucose, glutamine, and no pyruvate, Gibco®) to have 1×10^5 cells/mL. *B. abortus* 544 was grown in 2YT at 37 °C until exponential phase. The OD of the bacterial culture was measured, and dilutions were performed to have a MOI equal to 50 (50 times more bacteria than macrophages). An input control was performed for each condition by plating bacteria on a 2YT agar plate before infecting cells.

Cell medium was removed to add the appropriate bacterial dilution. The mix was centrifuged for 10 min at 1200 rpm (4 °C) and then incubated at 37 °C with 5% CO_2 (this time point is set as time zero). After one hour of incubation, medium was removed and replaced by medium containing gentamycin (50 µg/mL) for 1 h in order to kill extracellular bacteria, and then by medium containing gentamycin (10 µg/mL). Note that for the experiments using IPTG, the IPTG (10 mM) was kept during all the steps of the infection. At either 2 h, 4 h or 24 h post infection, cells were first washed with sterile PBS and were then incubated in PBS + Triton 0.1% at 37 °C for 10 min in order to lyse the cells while keeping bacteria alive. After that, cells were flushed and lysates were harvested. Serial dilutions were performed and each dilution was spotted on 2YT agar plates and incubated at 37 °C.

4.6. Infections of HeLa Cells

HeLa cells were plated in wells in DMEM medium (with sodium pyruvate, non-essential amino acid, glucose, glutamine, and no pyruvate, Gibco®) at 4 × 10^4 cells/mL. *B. abortus* 544 was grown in 2YT at 37 °C until exponential phase, the OD of the bacterial culture was measured, and dilutions were performed to have a MOI equal to 300. An input control was performed for each condition by plating bacteria on a 2YT agar plate before infecting cells. Prior to infections in the presence of IPTG (see below), *relA'* expression was induced 3 h before infection with IPTG (1 mM in YT medium). Cell medium was removed to add the appropriate bacterial dilution. The mix was centrifuged for 10 min at 1200 rpm (4 °C) and incubated at 37 °C with 5% CO_2 (this time point is set as time zero). After one hour of incubation, medium was removed and replaced by medium containing gentamycin (50 µg/mL) in order to kill extracellular bacteria, and then gentamycin (10 µg/mL). Note that for the experiments using IPTG, the IPTG (10 mM) was kept during all the steps of the infection. At either 2 h, 4 h or 24 h post infection, cells were first washed with sterile PBS and were then incubated in PBS + Triton 0.1% at 37 °C for 10 min in order to lyse the cells while keeping bacteria alive. After that, cells were flushed and lysates were harvested. Serial dilutions were performed and each dilution was spotted on 2YT agar plates and incubated at 37 °C.

4.7. G1 Counting

Bacteria in exponential phase of growth were diluted to an OD of 0.1 in 2YT liquid medium with or without IPTG. At each time point, 200 µL of the culture was washed two times in PBS and bacteria were loaded onto a PBS agarose pad to be observed and counted by fluorescence microscopy.

Supplementary Materials: The following are available online at http://www.mdpi.com/2076-0817/9/7/571/s1, Figure S1: Effect of dksA deletion on macrophage infection; Figure S2: Codon-adapted mesh1b sequence; Figure S3: Fluorescence microscopy of the pSRK-relA' mCherry-parB strain induced with IPTG; Table S1: Number of bacteria counted in Figure 5.

Author Contributions: M.V.d.H. and X.D.B. designed the work and wrote the manuscript; M.V.d.H. and E.C. performed the experiments. All authors have read and agreed to the published version of the manuscript.

Funding: This work was funded by FRS-FNRS (PDR T.0060.15 and T.0058.20) and the University of Namur.

Acknowledgments: We thank Maxime Quebatte (Biozentrum, Basel) for the generous gift of the pMQ203 plasmid. We thank Severin Ronneau for the generous gift of the pSRK-*relA'* plasmid and for the numerous fruitful discussions on (p)ppGpp. M.V.d.H. was supported by a FRIA Ph.D. fellowship from FRS-FNRS. We thank the University of Namur for logistic support.

Conflicts of Interest: The authors declare no competing interests.

References

1. Moreno, E.; Moriyon, I. The Genus *Brucella*. *Prokaryotes* **2006**, *5*, 315–456.
2. Deghelt, M.; Mullier, C.; Sternon, J.F.; Francis, N.; Laloux, G.; Dotreppe, D.; Van der Henst, C.; Jacobs-Wagner, C.; Letesson, J.J.; De Bolle, X. The newborn *Brucella abortus* blocked at the G1 stage of its cell cycle is the major infectious bacterial subpopulation. *Nat. Commun.* **2014**, *5*, 4366. [CrossRef] [PubMed]

3. Starr, T.; Ng, T.W.; Wehrly, T.D.; Knodler, L.A.; Celli, J. *Brucella* intracellular replication requires trafficking through the late endosomal/lysosomal compartment. *Traffic* **2008**, *9*, 678–694. [CrossRef] [PubMed]
4. Porte, F.; Liautard, J.P.; Kohler, S. Early acidification of phagosomes containing *Brucella suis* is essential for intracellular survival in murine macrophages. *Infect. Immun.* **1999**, *67*, 4041–4047. [CrossRef] [PubMed]
5. Pizarro-Cerda, J.; Meresse, S.; Parton, R.G.; van der Goot, G.; Sola-Landa, A.; Lopez-Goni, I.; Moreno, E.; Gorvel, J.P. *Brucella abortus* transits through the autophagic pathway and replicates in the endoplasmic reticulum of nonprofessional phagocytes. *Infect. Immun.* **1998**, *66*, 5711–5724. [CrossRef]
6. Starr, T.; Child, R.; Wehrly, T.D.; Hansen, B.; Hwang, S.; Lopez-Otin, C.; Virgin, H.W.; Celli, J. Selective subversion of autophagy complexes facilitates completion of the Brucella intracellular cycle. *Cell Host Microbe* **2012**, *11*, 33–45. [CrossRef] [PubMed]
7. Roop, R.M., 2nd; Gaines, J.M.; Anderson, E.S.; Caswell, C.C.; Martin, D.W. Survival of the fittest: How Brucella strains adapt to their intracellular niche in the host. *Med. Microbiol. Immunol.* **2009**, *198*, 221–238. [CrossRef]
8. Poncin, K.; Roba, A.; Jimmidi, R.; Potemberg, G.; Fioravanti, A.; Francis, N.; Willemart, K.; Zeippen, N.; Machelart, A.; Biondi, E.G.; et al. Occurrence and repair of alkylating stress in the intracellular pathogen *Brucella abortus*. *Nat. Commun.* **2019**, *10*, 4847. [CrossRef]
9. Kohler, S.; Foulongne, V.; Ouahrani-Bettache, S.; Bourg, G.; Teyssier, J.; Ramuz, M.; Liautard, J.P. The analysis of the intramacrophagic virulome of *Brucella suis* deciphers the environment encountered by the pathogen inside the macrophage host cell. *Proc. Natl. Acad. Sci. USA* **2002**, *99*, 15711–15716. [CrossRef]
10. Dozot, M.; Boigegrain, R.A.; Delrue, R.M.; Hallez, R.; Ouahrani-Bettache, S.; Danese, I.; Letesson, J.J.; De Bolle, X.; Kohler, S. The stringent response mediator Rsh is required for *Brucella melitensis* and *Brucella suis* virulence, and for expression of the type IV secretion system virB. *Cell Microbiol.* **2006**, *8*, 1791–1802. [CrossRef]
11. Kim, S.; Watanabe, K.; Suzuki, H.; Watarai, M. Roles of *Brucella abortus* SpoT in morphological differentiation and intramacrophagic replication. *Microbiology* **2005**, *151*, 1607–1617. [CrossRef] [PubMed]
12. Gonzalez, D.; Collier, J. Effects of (p)ppGpp on the progression of the cell cycle of *Caulobacter crescentus*. *J. Bacteriol.* **2014**, *196*, 2514–2525. [CrossRef] [PubMed]
13. Lesley, J.A.; Shapiro, L. SpoT regulates DnaA stability and initiation of DNA replication in carbon-starved *Caulobacter crescentus*. *J. Bacteriol.* **2008**, *190*, 6867–6880. [CrossRef] [PubMed]
14. Ronneau, S.; Petit, K.; De Bolle, X.; Hallez, R. Phosphotransferase-dependent accumulation of (p)ppGpp in response to glutamine deprivation in *Caulobacter crescentus*. *Nat. Commun.* **2016**, *7*, 11423. [CrossRef] [PubMed]
15. Schreiber, G.; Ron, E.Z.; Glaser, G. ppGpp-mediated regulation of DNA replication and cell division in *Escherichia coli*. *Curr. Microbiol.* **1995**, *30*, 27–32. [CrossRef]
16. Dalebroux, Z.D.; Edwards, R.L.; Swanson, M.S. SpoT governs *Legionella pneumophila* differentiation in host macrophages. *Mol. Microbiol.* **2009**, *71*, 640–658. [CrossRef]
17. Das, B.; Pal, R.R.; Bag, S.; Bhadra, R.K. Stringent response in *Vibrio cholerae*: Genetic analysis of *spoT* gene function and identification of a novel (p)ppGpp synthetase gene. *Mol. Microbiol.* **2009**, *72*, 380–398. [CrossRef]
18. Dahl, J.L.; Kraus, C.N.; Boshoff, H.I.; Doan, B.; Foley, K.; Avarbock, D.; Kaplan, G.; Mizrahi, V.; Rubin, H.; Barry, C.E., 3rd. The role of RelMtb-mediated adaptation to stationary phase in long-term persistence of *Mycobacterium tuberculosis* in mice. *Proc. Natl. Acad. Sci. USA* **2003**, *100*, 10026–10031. [CrossRef]
19. Paul, B.J.; Barker, M.M.; Ross, W.; Schneider, D.A.; Webb, C.; Foster, J.W.; Gourse, R.L. DksA: A critical component of the transcription initiation machinery that potentiates the regulation of rRNA promoters by ppGpp and the initiating NTP. *Cell* **2004**, *118*, 311–322. [CrossRef]
20. Ross, W.; Sanchez-Vazquez, P.; Chen, A.Y.; Lee, J.H.; Burgos, H.L.; Gourse, R.L. ppGpp Binding to a Site at the RNAP-DksA Interface Accounts for Its Dramatic Effects on Transcription Initiation during the Stringent Response. *Mol. Cell* **2016**, *62*, 811–823. [CrossRef]
21. Ronneau, S.; Hallez, R. Make and break the alarmone: Regulation of (p)ppGpp synthetase/hydrolase enzymes in bacteria. *FEMS Microbiol. Rev.* **2019**, *43*, 389–400. [CrossRef]
22. Mittenhuber, G. Comparative genomics and evolution of genes encoding bacterial (p)ppGpp synthetases/hydrolases (the Rel, RelA and SpoT proteins). *J. Mol. Microbiol. Biotechnol.* **2001**, *3*, 585–600. [PubMed]

23. Plommet, M. Minimal requirements for growth of *Brucella suis* and other *Brucella* species. *Zentralblatt Bakteriologie* **1991**, *275*, 436–450. [CrossRef]
24. Ronneau, S.; Caballero-Montes, J.; Coppine, J.; Mayard, A.; Garcia-Pino, A.; Hallez, R. Regulation of (p)ppGpp hydrolysis by a conserved archetypal regulatory domain. *Nucleic Acids Res.* **2019**, *47*, 843–854. [CrossRef]
25. Sternon, J.F.; Godessart, P.; Goncalves de Freitas, R.; Van der Henst, M.; Poncin, K.; Francis, N.; Willemart, K.; Christen, M.; Christen, B.; Letesson, J.J.; et al. Transposon Sequencing of *Brucella abortus* Uncovers Essential Genes for Growth In Vitro and Inside Macrophages. *Infect. Immun.* **2018**, *86*. [CrossRef]
26. Sun, D.; Lee, G.; Lee, J.H.; Kim, H.Y.; Rhee, H.W.; Park, S.Y.; Kim, K.J.; Kim, Y.; Kim, B.Y.; Hong, J.I.; et al. A metazoan ortholog of SpoT hydrolyzes ppGpp and functions in starvation responses. *Nat. Struct. Mol. Biol.* **2010**, *17*, 1188–1194. [CrossRef]
27. Khan, S.R.; Gaines, J.; Roop, R.M., 2nd; Farrand, S.K. Broad-host-range expression vectors with tightly regulated promoters and their use to examine the influence of TraR and TraM expression on Ti plasmid quorum sensing. *Appl. Environ. Microbiol.* **2008**, *74*, 5053–5062. [CrossRef]
28. Hanna, N.; Ouahrani-Bettache, S.; Drake, K.L.; Adams, L.G.; Kohler, S.; Occhialini, A. Global Rsh-dependent transcription profile of *Brucella suis* during stringent response unravels adaptation to nutrient starvation and cross-talk with other stress responses. *BMC Genom.* **2013**, *14*, 459. [CrossRef]
29. Dozot, M.; Poncet, S.; Nicolas, C.; Copin, R.; Bouraoui, H.; Maze, A.; Deutscher, J.; De Bolle, X.; Letesson, J.J. Functional characterization of the incomplete phosphotransferase system (PTS) of the intracellular pathogen *Brucella melitensis*. *PLoS ONE* **2010**, *5*. [CrossRef]
30. Martinez-Nunez, C.; Altamirano-Silva, P.; Alvarado-Guillen, F.; Moreno, E.; Guzman-Verri, C.; Chaves-Olarte, E. The two-component system BvrR/BvrS regulates the expression of the type IV secretion system VirB in *Brucella abortus*. *J. Bacteriol.* **2010**, *192*, 5603–5608. [CrossRef]
31. Lacerda, T.L.; Salcedo, S.P.; Gorvel, J.P. Brucella T4SS: The VIP pass inside host cells. *Curr. Opin. Microbiol.* **2013**, *16*, 45–51. [CrossRef] [PubMed]
32. Terwagne, M.; Mirabella, A.; Lemaire, J.; Deschamps, C.; De Bolle, X.; Letesson, J.J. Quorum sensing and self-quorum quenching in the intracellular pathogen *Brucella melitensis*. *PLoS ONE* **2013**, *8*, e82514. [CrossRef] [PubMed]
33. Choi, K.H.; Gaynor, J.B.; White, K.G.; Lopez, C.; Bosio, C.M.; Karkhoff-Schweizer, R.R.; Schweizer, H.P. A Tn7-based broad-range bacterial cloning and expression system. *Nat. Methods* **2005**, *2*, 443–448. [CrossRef] [PubMed]

pathogens

MDPI

Article

Brucella abortus-Stimulated Platelets Activate Brain Microvascular Endothelial Cells Increasing Cell Transmigration through the Erk1/2 Pathway

Ana María Rodríguez [1,†], Aldana Trotta [2,†], Agustina P. Melnyczajko [1], M. Cruz Miraglia [1,§], Kwang Sik Kim [3], M. Victoria Delpino [1], Paula Barrionuevo [2,‡] and Guillermo Hernán Giambartolomei [1,*,‡]

1 Instituto de Inmunología, Genética y Metabolismo (INIGEM), CONICET, Facultad de Farmacia y Bioquímica, Universidad de Buenos Aires, Buenos Aires C1120AAD, Argentina; amrodriguez@fmed.uba.ar (A.M.R.); amelnyczajko@uade.edu.ar (A.P.M.); miraglia.maria@inta.gob.ar (M.C.M.); mdelpino@ffyb.uba.ar (M.V.D.)

2 Instituto de Medicina Experimental (IMEX) (CONICET-Academia Nacional de Medicina), Buenos Aires C1425ASU, Argentina; aldana.trotta@unahur.edu.ar (A.T.); pbarrionuevo@fbmc.fcen.uba.ar (P.B.)

3 Division of Pediatric Infectious Diseases, Department of Pediatrics, Johns Hopkins University School of Medicine, Baltimore, MD 21287, USA; kwangkim@jhmi.edu

* Correspondence: ggiambart@ffyb.uba.ar

† These authors have contributed equally to this work.

‡ These authors share senior authorship.

§ Present address: Instituto de Virología, Centro de Investigación en Ciencias Veterinarias y Agronómicas, Instituto Nacional de Tecnología Agropecuaria (INTA), Hurlingham B1686, Argentina.

Received: 27 July 2020; Accepted: 25 August 2020; Published: 27 August 2020

Abstract: Central nervous system invasion by bacteria of the genus *Brucella* results in an inflammatory disorder called neurobrucellosis. A common feature associated with this pathology is blood–brain barrier (BBB) activation. However, the underlying mechanisms involved with such BBB activation remain unknown. The aim of this work was to investigate the role of *Brucella abortus*-stimulated platelets on human brain microvascular endothelial cell (HBMEC) activation. Platelets enhanced HBMEC activation in response to *B. abortus* infection. Furthermore, supernatants from *B. abortus*-stimulated platelets also activated brain endothelial cells, inducing increased secretion of IL-6, IL-8, CCL-2 as well as ICAM-1 and CD40 upregulation on HBMEC compared with supernatants from unstimulated platelets. Outer membrane protein 19, a *B. abortus* lipoprotein, recapitulated *B. abortus*-mediated activation of HBMECs by platelets. In addition, supernatants from *B. abortus*-activated platelets promoted transendothelial migration of neutrophils and monocytes. Finally, using a pharmacological inhibitor, we demonstrated that the Erk1/2 pathway is involved in the endothelial activation induced by *B. abortus*-stimulated platelets and also in transendothelial migration of neutrophils. These results describe a mechanism whereby *B. abortus*-stimulated platelets induce endothelial cell activation, promoting neutrophils and monocytes to traverse the BBB probably contributing to the inflammatory pathology of neurobrucellosis.

Keywords: *Brucella abortus*; neurobrucellosis; platelets; brain microvascular endothelial cells; endothelial cells

1. Introduction

Blood–brain barrier (BBB) integrity is necessary to protect the brain from injuries such as toxins and germs, as well as to help in maintaining central nervous system (CNS) homeostasis [1]. BBB activation and dysfunction contributes to several brain pathologies. Many factors are able to induce BBB dysfunction such as inflammatory mediators, matrix metalloproteinases, free radicals, and vascular endothelial growth factor, among others [2].

Bacteria of the genus *Brucella* produce several types of inflammatory disorders [3]. Neurobrucellosis is a neurodegenerative inflammatory disorder caused by invasion of the CNS by *Brucella* spp. and constitutes the most morbid pathology associated with this infection [4]. One of the most characteristic clinical signs of this disease is pleocytosis; i.e., the presence of leukocytes in the cerebrospinal fluid [4].

We have recently described a putative mechanism employed by *Brucella abortus* to gain access to the CNS. We have demonstrated, using an in vitro model, that *B. abortus* traverses the BBB into the cerebral parenchyma inside infected monocytes, by a mechanism known as "Trojan horse". Moreover, infected monocytes act as bacterial source for de novo infection of glial cells [5]. In addition, we have described that activation of glial cells by *B. abortus* is crucial in neurobrucellosis pathology [6]. *B. abortus*-activated astrocytes and microglia secrete pro-inflammatory mediators such as tumor necrosis factor (TNF)-α, interleukin (IL)-6, IL-1β, C-C motif chemokine ligand 2 CCL-2, C-X-C motif chemokine ligand 1 (CXCL1), metalloproteinase (MMP)-9, and nitric oxide (NO), among others [6–8]. These inflammatory mediators, and astrocytes/microglia-secreted IL-1β in particular, activate the BBB, allowing monocyte and neutrophil transmigration [9]. The effect of glial cell activation on BBB cells is well known [9–11]. However, whether peripheral inflammation induced by *Brucella*-activated cells can modify the BBB remains unknown.

Platelets are well characterized as responsible for maintaining vascular integrity in addition to being hemostatic mediators [12]. In the last few years, the immune function of platelets has been described in both homeostasis and pathology [12–14]. Platelets express several immune receptors such as toll-like receptors (TLR) and Fc receptors, which allow the recognition of different pathogens [12]. Upon pathogen recognition, platelets can be activated and secrete a wide variety of immunomodulatory mediators present in their granules [15–17]. The immunoregulatory functions of pathogen-activated platelets have been described recently, as well as their ability to activate endothelial cells, including the microvascular endothelium of the BBB [14,18–20]. We have recently described the interactions between platelets and *B. abortus* [14]. *B. abortus* is able to invade, infect platelets, and activate them. As a consequence of this interaction, platelets establish complexes with *B. abortus*-infected monocytes, increasing the efficiency of the infection and modulating monocyte and neutrophil functions [14,21]. Moreover, we demonstrated that complexes between platelets and both monocytes and neutrophils are more abundant in patients with brucellosis than in healthy donors [21].

We have already described the effect of glial cell activation on BBB cells during neurobrucellosis; however, whether platelets activated by *Brucella* can modify the BBB remains unexplored. Therefore, the aim of this work was to elucidate whether *B. abortus*-activated platelets can activate the BBB and affect transmigration of monocytes and neutrophils. Here, we demonstrated that *B. abortus*-activated platelets activate brain microvascular endothelial cells, and other endothelial cells, through Erk1/2 signaling pathway, leading monocytes and neutrophils to traverse polarized brain microvascular endothelial monolayers.

2. Results

2.1. *Interaction with Platelets Enhances the Activation of Endothelial Cells in the Context of* B. abortus *Infection*

We decided to evaluate the capacity of *B. abortus* to activate human brain microvascular endothelial cells (HBMECs) in presence of platelets. For this, HBMECs were co-cultured with platelets (cell:platelet ratio, 1:100) and infected with *B. abortus* (MOI of 100) for 24 h. As control, HBMECs were cultured with platelets alone or they were infected in the absence of platelets. We measured ICAM-1 (intercellular adhesion molecule 1, also known as CD54) expression to determine the level of activation of endothelial cells. Endothelial ICAM-1 plays a critical role at different steps of neutrophil migration into inflamed tissues [22]. As we have previously reported, infection with *B. abortus* in the absence of platelets induced a slight activation of HBMECs, measured as the upregulation of surface ICAM-1 [9]. The presence of platelets in the absence of infection also induced a slight activation of HBMECs. However, these effects were not significant. On the other hand, the presence of platelets during the infection of HBMECs induced a significant ($p < 0.0005$) upregulation of ICAM-1 (Figure 1A). These results demonstrate that the presence of platelets enhances *B. abortus*-induced activation of microvascular brain endothelial cells.

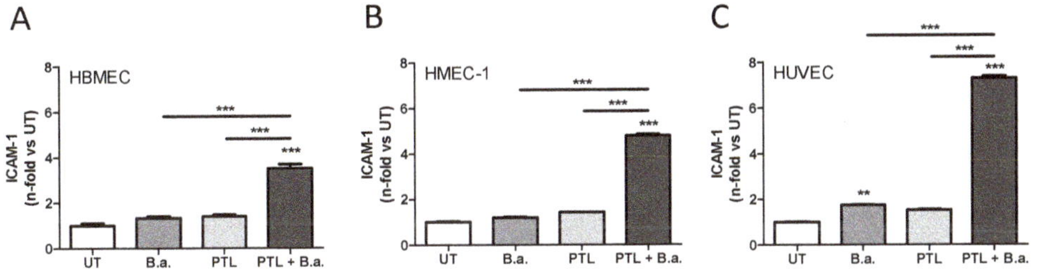

Figure 1. Interaction with platelets enhances the activation of endothelial cells in the context of *Brucella abortus* infection. Endothelial cells lines human brain microvascular endothelial cell (HBMEC) (**A**), human microvascular endothelial cells (HMEC-1) (**B**), and human umbilical vein endothelial cell (HUVEC) primary culture (**C**) were incubated with *B. abortus* (B.a.), in the presence or the absence of platelets (PTL), or with PTL alone for 24 h. ICAM-1 (intercellular adhesion molecule 1) expression on the cell surface was assessed by flow cytometry. Bars represent the mean ± SEM of duplicates. Data shown are from a representative experiment out of at least three performed. ** $p < 0.005$, *** $p < 0.0005$ vs. untreated cells (UT).

To investigate whether the effect of platelets during *B. abortus* infection also occurs in other endothelia, human microvascular endothelial cells (HMEC-1) and human umbilical vein endothelial cell (HUVEC) were infected with *B. abortus* in the presence or absence of platelets. The presence of platelets during the infection of both types of endothelial cells induced a significant upregulation of ICAM-1 expression compared to infected cells or cells incubated with platelets alone ($p < 0.0005$) (Figure 1B,C). These data demonstrate that the presence of platelets enhances *B. abortus*-activation of different endothelial cell types.

2.2. Supernatants from B. abortus-Stimulated Platelets Activate Brain Microvascular Endothelial Cells

To investigate whether this effect was due to direct interaction between platelets and endothelial cells or factors released by *B. abortus*-activated platelets, we performed experiments using conditioned media. First, platelets were stimulated with or without *B. abortus* (platelets: *B. abortus* ratio, 1:1) for 24 h. Then, culture supernatants were collected and filtered to eliminate platelets and bacteria. Finally, cell-free supernatants were used to stimulate HBMECs for an additional 24 h. Stimulation of HBMECs with supernatants from *B. abortus*-stimulated platelets induced a significant ($p < 0.005$) upregulation of ICAM-1 surface expression (Figure 2A). These results demonstrate that supernatants from *B. abortus*-stimulated platelets are able to activate microvascular brain endothelial cells. Furthermore, in order to expand our results, we investigated whether these supernatants could also activate other endothelial cell types. For this, HMEC-1 and HUVEC were stimulated with supernatants collected from *B. abortus*-stimulated platelets. We observed an upregulation of ICAM-1 surface expression on both cell types (Figure 2B,C). Collectively, these data demonstrated that supernatants from *B. abortus*-stimulated platelet are able to activate several types of endothelial cells.

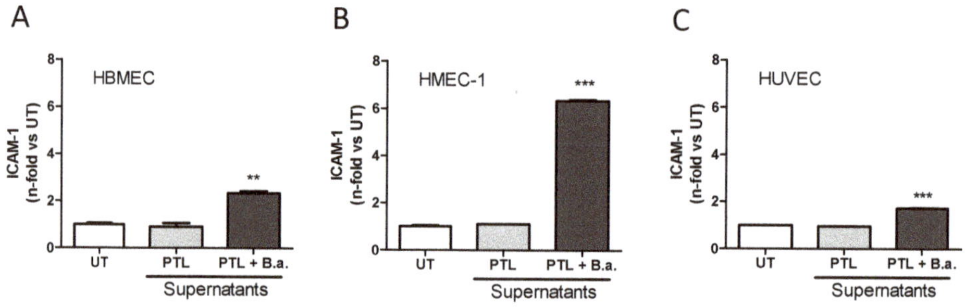

Figure 2. Supernatants from *B. abortus*-stimulated platelets activate endothelial cells. Supernatants collected from *B. abortus*-stimulated platelets (PTL + B.a.) or platelets alone (PTL) were used to stimulate the endothelial cell lines HBMEC (**A**), HMEC-1 (**B**), and HUVEC primary culture (**C**) for 24 h (dilution 1:2). ICAM-1 expression was measured on the cell surface by flow cytometry. ** $p < 0.005$, *** $p < 0.0001$ vs. untreated cells (UT).

Next, we studied in depth the HBMEC activation induced by supernatants from *B. abortus*-stimulated platelets. Activation of HBMECs induced by supernatants from *B. abortus*- stimulated platelets was dose dependent (Figure 3A) and, more importantly, it was achieved by using supernatants from different platelet donors (Figure 3B). CD40 expression in endothelial cells has been implicated in several pathologic conditions of the CNS including Alzheimer's disease and human immunodeficiency virus 1 (HIV-1) encephalitis, where an important role of CD40 has been demonstrated in BBB disruption [23]. Besides the upregulation of ICAM-1, the activated phenotype induced by *B. abortus*-infected platelet supernatants also included the significant upregulation of CD40 surface expression ($p < 0.05$) (Figure 3C) and the secretion of significant ($p < 0.0005$) amounts of IL-6, IL-8, and CCL-2 (Figure 3D–F) when compared with untreated HBMECs or HBMECs treated with supernatants from unstimulated platelets. Levels of activation were comparable to those obtained when HBMECs were stimulated with culture supernatants from *Brucella*-infected astrocytes [9] or IL-1β used as positive controls. Importantly, the concentrations of IL-6, IL-8, and CCL-2 measured in supernatants from *B. abortus*-stimulated platelets used for stimulation were negligible (<200 pg/mL in all cases, data not shown). These results demonstrate that supernatants from *B. abortus*-stimulated platelets induce an activated phenotype in microvascular brain endothelial cells, characterized by the upregulation of surface molecules such as ICAM-1 and CD40, and the secretion of both cytokines and chemokines.

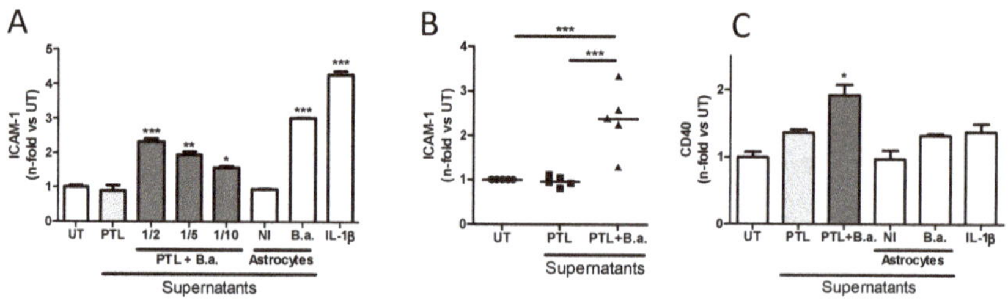

Figure 3. *Cont.*

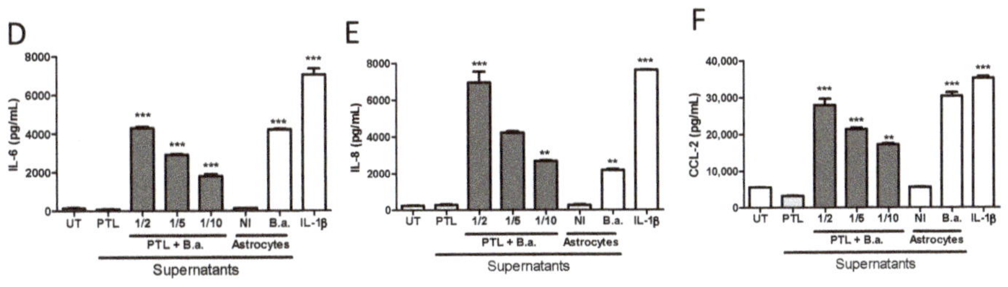

Figure 3. Supernatants from *B. abortus*-infected platelets induce an activated phenotype in HBMECs. Platelets were incubated with *B. abortus* (PTL + B.a.) or alone (PTL) for 24 h. Culture supernatants from *B. abortus*-infected astrocytes and interleukin (IL)-1β were used as control. Platelets supernatants were collected, filtered, and used to stimulate HBMECs for 24 h at the indicated dilutions. (**A**) ICAM-1 expression was measured by flow cytometry on HBMEC surface. (**B**) ICAM-1 expression obtained by stimulating HBMECs with supernatants obtained from five independent PTL donors. (**C**) CD40 expression was measured on HBMEC surface by flow cytometry. The secretion of IL-6 (**D**), IL-8 (**E**), and C-C motif chemokine ligand 2 (CCL-2) (**F**) was determined from HBMEC treated culture supernatants by ELISA. Bars represent the mean ± SEM of duplicates. Data shown are from a representative experiment out of at least three performed, except B. * $p < 0.05$, ** $p < 0.005$, *** $p < 0.0005$ vs. untreated cells (UT). NI: non infected.

2.3. Secreted Factors from B. abortus-Activated Platelets Activate HBMECs

Previous studies have shown that *Brucella* spp. release outer-membrane vesicles (OMVs, also known as blebs) containing lipopolysaccharide (LPS), outer membrane proteins, and other bacterial components [24]. To rule out the possibility that OMVs were implicated in HBMEC activation, they were removed from the supernatants by ultracentrifugation, as previously described [24]. HBMECs were then incubated with OMVs-free supernatants for 24 h and the activation of HBMECs was evaluated. There were no significant differences ($p > 0.05$) between non-depleted and OMVs-free supernatants regarding HBMEC activation, measured as ICAM-1 expression (Figure 4A) and IL-6, IL-8, and CCL-2 secretion (Figure 4B–D, respectively). To discard any putative participation of *Brucella*-secreted factors on HBMEC activation, *B. abortus* was incubated alone in the same culture conditions for 24 h. Then, culture supernatants were filtered and ultracentrifuged as described above. These platelet-free *B. abortus* culture supernatants were unable to activate HBMECs (Figure 4). Altogether, these results indicate that secreted factors from *B. abortus*-activated platelets are responsible for HBMEC activation.

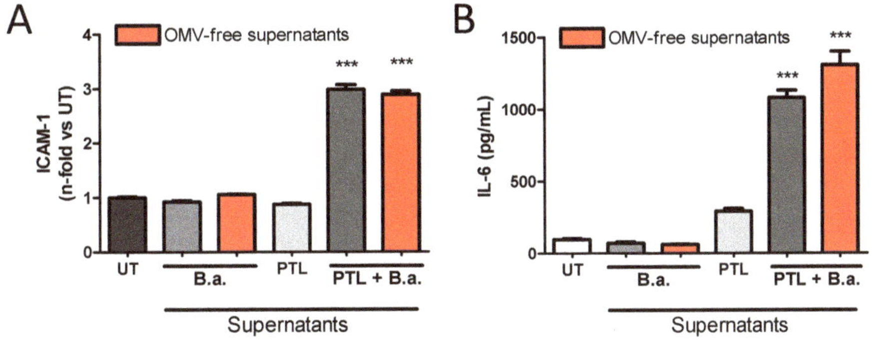

Figure 4. *Cont.*

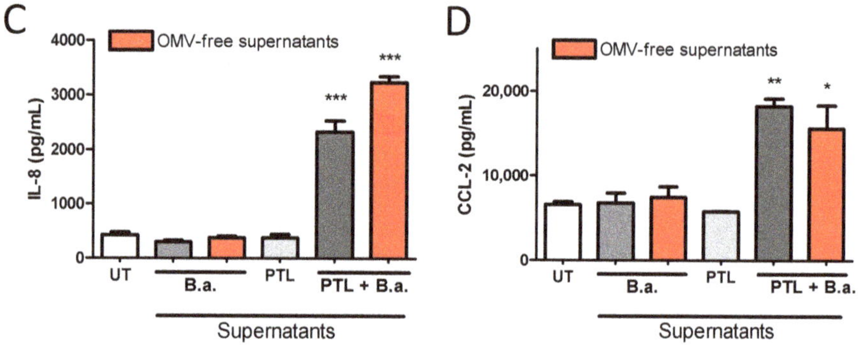

Figure 4. Secreted factors from *B. abortus*-stimulated platelets activate HBMECs. Supernatants from *B. abortus*-activated platelets (PLT + B.a.) or from *B. abortus* alone were ultracentrifuged (outer-membrane vesicle (OMV)-free supernatants) or not and used to stimulate HBMECs for 24 h. Supernatants from non-ultracentrifuged PTL alone were used as control. (**A**) ICAM-1 surface expression on HBMECs was assessed by flow cytometry. HBMEC secretion of IL-6 (**B**), IL-8 (**C**), and CCL-2 (**D**) was quantified by ELISA. Bars represent the mean ± SEM of duplicates from a representative experiment out of at least three performed. * $p < 0.05$, ** $p < 0.005$, *** $p < 0.0001$ vs. untreated cells (UT).

2.4. *B. abortus Lipoprotein-Stimulated Platelets Activate Brain Microvascular Endothelial Cells*

To test whether bacterial viability was necessary to induce the activation of platelets and consequently HBMEC activation, platelets were incubated for 24 h with heat-killed *B. abortus* (HKBA). Supernatants were then filtered and used as stimuli on HBMECs. HKBA supernatants were used to treat HBMECs as a negative control. As positive control, platelets were activated with thrombin. Supernatants from HKBA-stimulated platelets were able to activate HBMECs, inducing the upregulation of ICAM-1 (Figure 5A), and increasing the secretion of IL-6, IL-8, and CCL-2 (Figure 5B–D, respectively). We have previously demonstrated that different cell types can be activated by *B. abortus* lipoproteins [8,25,26]. Therefore, we further evaluated the contribution of lipoproteins in the induction of HBMEC activation by platelets. For this, platelets were incubated with *B. abortus* lipidated- or unlipidated-outer-membrane protein 19 (L-Omp19 or U-Omp19, respectively), used as a *Brucella* lipoprotein model [25]. Then, HBMECs were stimulated with the filtered supernatants for an additional 24 h, and the expression of ICAM-1 and secretion levels of IL-6, IL-8, and CCL-2 were evaluated. L-Omp19-activated platelets recapitulated HBMEC activation induced by supernatants from *B. abortus*-stimulated platelets. Furthermore, this activation was dependent on the lipidation of Omp19, as U-Omp19-stimulated platelets failed to induce HBMEC activation (Figure 5A–D). Culture supernatants from thrombin-activated platelets also induced partial activation of HBMECs. Neither HKBA-, L-Omp19-, nor U-Omp19-stimulated supernatants were able to activate HBMECs, demonstrating the presence of a platelet-secreted factor involved in the activation of HBMECs. Altogether, these results demonstrated that the presence of supernatants from platelets stimulated by structural components of *Brucella* (such as L-Omp19), independently of bacterial viability, are involved in the activation of HBMECs.

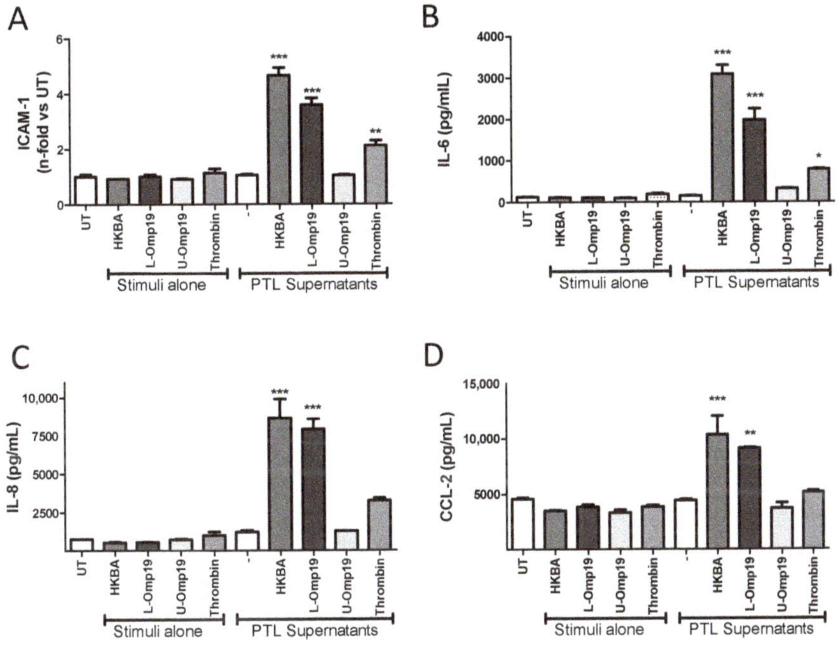

Figure 5. *B. abortus* lipoprotein-stimulated platelets activate HBMECs. Platelets were stimulated for 24 h with heat killed *B. abortus* (HKBA, 10^8 bacteria/mL), lipidated-outer-membrane protein 19 (L-Omp19), unlipidated-outer-membrane protein 19 (U-Omp19) (500 ng/mL), or thrombin (0.1 U/mL). Then, supernatants were collected, filtered. and used to stimulate HBMECs for an additional 24 h. Supernatants from PTL and stimuli incubated alone at the same conditions were used as control. ICAM-1 (**A**) was determined on HBMEC surface by flow cytometry. The secretion of IL-6 (**B**), IL-8 (**C**), and CCL-2 (**D**) was determined by ELISA. Bars represent the mean ± SEM of duplicates from a representative experiment out of at least three performed. * $p < 0.05$, ** $p < 0.005$, *** $p < 0.0001$ vs. untreated cells (UT).

2.5. Platelet-Stimulated HBMECs Induce Transendothelial Migration of Neutrophils and Monocytes

The presence of leukocytes in the cerebrospinal fluid and cerebral parenchyma has been described during neurobrucellosis [4]. This phenomenon, named pleocytosis, could be a consequence of the activation induced by *Brucella*-activated platelets on the blood–brain barrier [9]. To test this possibility, we used a previously established assay of transendothelial migration [9]. Briefly, HBMECs were seeded in the upper chamber of a Transwell plate, and they were cultured for 5 days to establish a monolayer. Then, HBMEC monolayers were treated for 24 h with supernatants from *B. abortus*-stimulated platelets. Culture supernatants from *Brucella*-infected astrocytes and human IL-1β were used as control. Finally, neutrophils or monocytes were seeded in the upper chamber and incubated for 3 h and the number of transmigrated cells to the lower chamber was quantified.

Monocyte as well as neutrophil migration increased when the HBMEC monolayer was treated with supernatants from *B. abortus*-stimulated platelets (Figure 6A,B), but not when HBMECs were stimulated with supernatants from platelets alone. Cellular transmigration was comparable to that obtained when HBMECs where stimulated with culture supernatants from *Brucella*-infected astrocytes or IL-1β used as positive controls. These results indicate that activation of brain endothelial cells by supernatants from *B. abortus*-stimulated platelets could induce transmigration of immune cells through a polarized brain endothelial cell monolayer. Taken together, these results suggest that activated platelets could be responsible for the induction of pleocytosis in the context of *B. abortus* CNS infection.

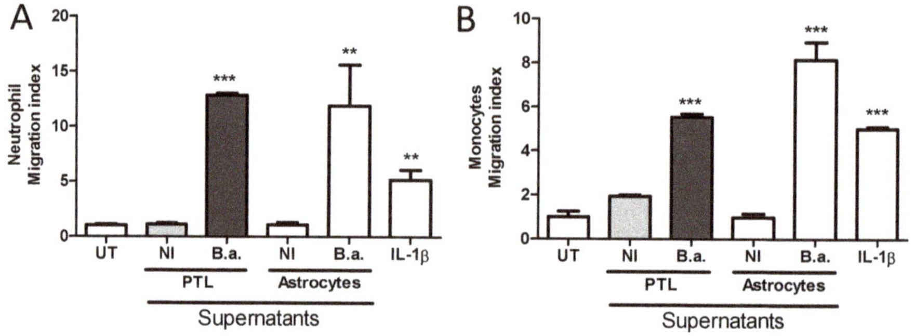

Figure 6. *B. abortus*-infected platelets activate HBMEC, promoting transendothelial migration of neutrophils and monocytes. HBMEC monolayers were established on the membrane of Transwell plates and then treated with supernatants from *B. abortus*-stimulated platelets for additional 24 h. Culture supernatants from *Brucella*-infected astrocytes and human IL-1β were used as control. Next, neutrophils (**A**) or monocytes (**B**) were seeded in the upper chamber and incubated for 3 h. Finally, media from the lower chamber were harvested and the number of migrated cells was quantified. Bars represent the mean ± SEM of duplicates from a representative experiment of at least three performed. ** $p < 0.005$, *** $p < 0.0005$ vs. untreated cells (UT) or indicated treatment. NI: non infected.

2.6. The Erk1/2 Pathway Is Involved in HMBEC Activation Induced by B. abortus-Stimulated Platelets and It Is Implicated in Transendothelial Migration of Neutrophils

We decided to further investigate the molecular mechanisms involved in endothelial activation and cellular transcytosis. It was previously reported that the extracellular signal-regulated kinase (Erk)1/2 pathway is involved in the activation of brain endothelial cells [27]. Taking this into account, we investigated the participation of the Erk1/2 signaling pathway in the activation of HBMECs by supernatants from *B. abortus*-stimulated platelets. For this, we used the Erk1/2-specific inhibitor PD98059 to treat HBMEC cells. Inhibition of the Erk1/2 pathway partially reduced the upregulation of ICAM-1 induced by supernatants from *B. abortus*-stimulated platelets ($p < 0.005$) (Figure 7A). Moreover, our results showed that cytokine and chemokine secretion of activated HBMECs is also regulated by the Erk1/2 pathway, since the inhibition with PD98059 also partially diminished the secretion of IL-6, IL-8, and CCL-2, compared to non-treated cells (Figure 7B–D). These results indicate that the Erk1/2 pathway is involved in HBMEC activation by supernatants from *B. abortus*-activated platelets.

The involvement of the Erk1/2 pathway in ICAM-1 upregulation is particularly interesting since ICAM-1 is one of the immunoglobulin-like cell adhesion molecules implicated in the transendothelial migration of immune cells [22]. Thus, we investigated the involvement of the Erk1/2 pathway on the increasing transendothelial migration throughout HBMECs activated by supernatants from *B. abortus*-stimulated platelets. For this, HBMEC monolayers on Transwells were pre-treated with PD98059 2 h before and throughout treatment with *B. abortus*-stimulated platelet supernatants. 24 h later, neutrophils were seeded in the upper chamber for 3 h and the number of migrated cells to the lower chamber was quantified. Neutrophil migration was totally inhibited in HBMECs pre-treated with PD98059, demonstrating the implication of the Erk1/2 pathway on the increased transmigration of immune cells (Figure 7E).

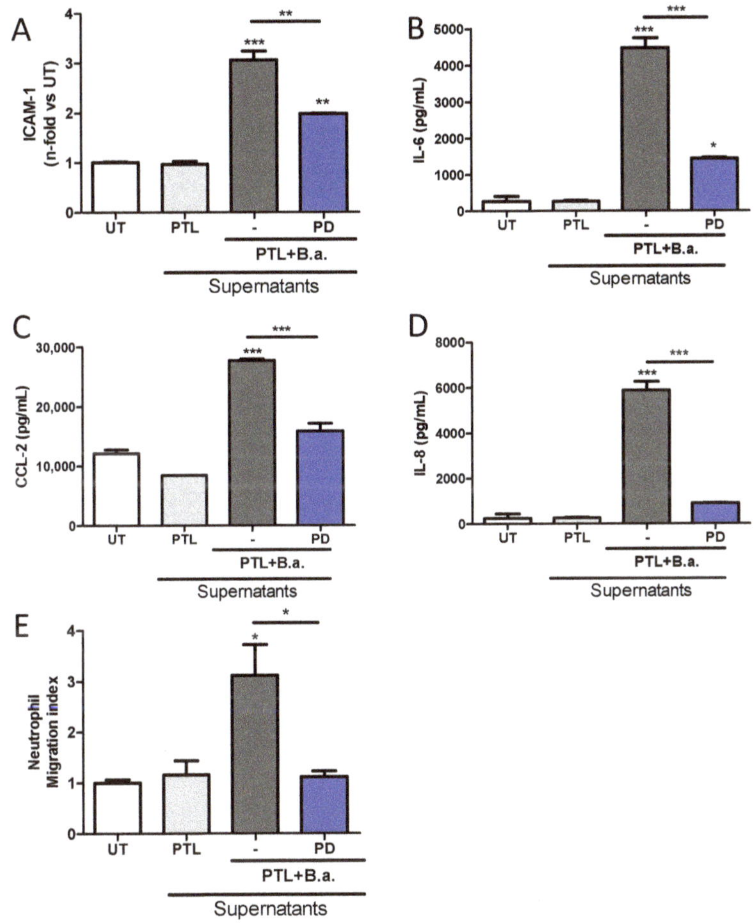

Figure 7. Extracellular signal-regulated kinase (Erk)1/2 pathway is involved in HBMEC activation induced by *B. abortus*-stimulated platelets and it is implicated in transendothelial migration of neutrophils. HBMEC were pre-incubated with the Erk1/2 inhibitor (PD98059) for 2 h before platelets–supernatants stimulation and kept throughout. ICAM-1 was determined on HBMECs surface by flow cytometry (**A**). The secretion of IL-6 (**B**), IL-8 (**C**), and CCL-2 (**D**) was determined by ELISA. HBMEC monolayers were established on the membrane of Transwell plates. HBMEC activation was inhibited by PD98059 and then treated with supernatants from *B. abortus*-stimulated platelets for additional 24 h. Next, neutrophils were seeded in the upper chamber and incubated for 3 h. Finally, media from the lower chamber was harvested and the number of migrated cells was quantified (**E**). Bars represent the mean ± SEM of duplicates from a representative experiment of at least three performed * $p < 0.05$, ** $p < 0.005$, *** $p < 0.0005$ vs. untreated cells (UT) or indicated treatment.

3. Discussion

Physiological and pathological immune responses are a continuum in which platelets are recognized as innate immune effector cells. Their activation stimulates interactions with endothelial cells and myeloid leukocytes in many pathologic inflammatory syndromes, as well as consequences in acute inflammation [28–31]. Platelets also have signaling functions in endothelial cells. These functions also contribute to critical inflammatory and immune responses [32,33].

Brain microvascular endothelium activation and BBB dysfunction is a significant contributor to the pathogenesis of a variety of brain pathologies [2], many of them of microbial origin [9,18,34]. We have previously described the ability of *B. abortus* to induce inflammation in the cerebral parenchyma, which leads to the activation of the endothelial cells that form the BBB [9]. In this study, we elucidated the role of platelets in brain microvascular endothelial cell activation mediated by *B. abortus*. Platelets enhance HBMEC activation in the context of *B. abortus* infection. These results correlate with the reported ability of other bacterial species to activate platelets and harm endothelial cells [35,36].

Interestingly, HBMEC activation does not require direct contact between platelets and brain endothelial cells, since supernatants of *B. abortus*-stimulated platelets recapitulated the HBMEC activation observed in the presence of platelets. Furthermore, HBMEC activation by secreted factors from *B. abortus*-stimulated platelets is sufficient to induce transmigration of both monocytes and neutrophils. Moreover, *B. abortus*-stimulated platelets also activate the HMEC-1 cell line and primary culture of HUVEC, underscoring the ability of *B. abortus*-stimulated platelets to activate any endothelium.

A long time ago, it was demonstrated that activated platelets increase CCL-2 secretion and ICAM-1 expression on HUVECs [37]. This indicates that activated platelets are able to change the chemotactic and adhesive properties of endothelial cells, increasing the ability to attract monocytes and neutrophils. Under physiological conditions, endothelial cells of the vasculature of non-inflamed tissues have as main functions the maintenance of blood fluidity and the control of vascular permeability [33]. Under these conditions, resting endothelial cells do not interact with circulating leukocytes since the proteins necessary for this interaction are mainly retained inside the cell [38]. Under acute inflammatory conditions, such as those induced by *B. abortus* infection, the vascular endothelium is rapidly activated, mobilizing these adhesion molecules to the extracellular membrane [33]. In accordance with this, we demonstrated that, although the infection with *B. abortus* induces a mild activation of HBMEC, HMEC-1, and HUVEC, the presence of platelets during the infection enhances its activation state upregulating the expression of ICAM-1 and CD40, thus stressing the amplifying role of platelets on endothelial inflammation [39]. In line with these results, other authors have shown that the presence of activated platelets significantly induces the expression of E-Selectin (CD62E), CD106 (VCAM-1), and ICAM-1 on the surface of HUVEC cells, even in the absence of others inflammatory agents [40]. In addition to the increase in adhesion molecules, we have demonstrated that the activation of HBMECs by supernatants from *B. abortus*-stimulated platelets increase IL-6, IL-8, and CCL-2 secretion. These results are in agreement with those previously published describing that HUVEC secrete IL-8 and CCL-2 after co-incubation with activated platelets [40]. In turn, in vivo experiments have shown that platelets are one of the first cellular components arrested in the inflamed endothelium, promoting their activation and allowing the subsequent arrest of leukocytes [41].

Platelet activation was also induced by exposure to heat-killed *B. abortus*, which indicated that it was not dependent on bacterial viability and suggests that it was elicited by a structural bacterial component. Our laboratory has been investigating for years the role of lipoproteins in inflammation generated by *Brucella*. We have described that *Brucella* LPS does not produce cellular activation, however, *Brucella* lipoproteins produce activation of several cell types [6,8,25,26]. Thus, we hypothesized that *B. abortus* lipoproteins might be the structural components involved in the observed phenomenon. L-Omp19, a prototypical *B. abortus* lipoprotein, recapitulated platelet stimulation and concomitant HBMEC activation. Acylation of Omp19 was required for its biological activity since U-Omp19 had no effect on platelet stimulation. The genome of *B. abortus* possesses no less than 80 genes encoding putative lipoproteins [42], and many of them are expressed in the outer membrane of the bacterium [43]. In this context, we posit that any surface-exposed *Brucella* lipoprotein may be significant beyond in vitro assays and not one lipoprotein but rather a combination of them may contribute to the platelet activation elicited by *B. abortus*.

Our recent work revealed a physiological mechanism employed by *B. abortus* to traverse the BBB. *Brucella* is incapable of traversing the BBB by itself, despite the ability to invade and replicate in endothelial cells of the brain microvasculature. Instead, it could cross a BBB model in vitro as a consequence of naturally migrating monocytes carrying viable bacteria, which serve as source of de novo infection to astrocytes and microglia [5]. Interestingly, we have also demonstrated that activated *B. abortus*-infected glial cells were able to increase the transmigration of monocytes through the secretion of inflammatory mediators [9]. These mediators would escalate the entering of infected cells from the peripheral circulation, increasing the infection and the subsequent BBB dysfunction through a pathological vicious circle. The capacity of secreted factors from *B. abortus*-stimulated platelets to increase neutrophil and monocyte transmigration through microvascular endothelial cells demonstrated in this paper would worsen this situation (Figure 8).

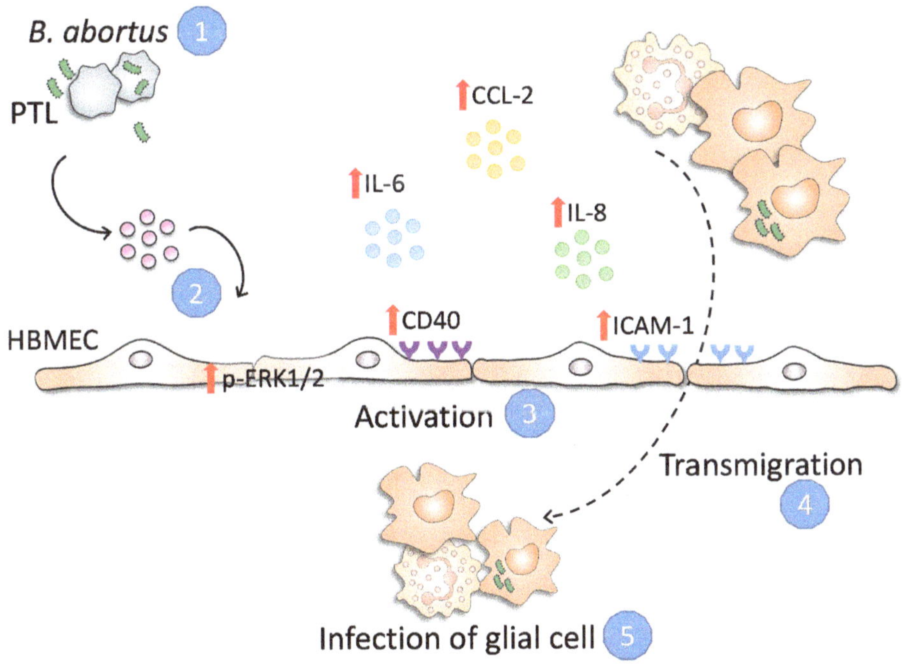

Figure 8. *B. abortus*-stimulated platelets (1) secret factors (2) that induce HBMEC activation, leading to ICAM-1 and CD40 upregulation, increasing the secretion of IL-6, IL-8, and CCL-2 (3), and promoting neutrophils and monocytes to traverse a polarized HMBEC monolayers (4). Platelet-induced activation would escalate the entering of infected cells from the peripheral circulation and the subsequent infection of glial cells (5), worsening the inflammatory signs of neurobrucellosis.

The mitogen-activated protein kinase (MAPK) pathway has been associated to several biological processes such as cell activation and proliferation, cell differentiation, and apoptosis [44]. In particular, the Erk1/2 pathway is involved in HBMEC activation [27] and endothelial permeability [45]. Experiments of pharmacological inhibition determined that Erk1/2 was involved in HBMEC activation induced by supernatants from *B. abortus*-activated platelets. In particular, it was involved in ICAM-1 upregulation and enhanced the transmigration of neutrophils. Since MAPK inhibitors, such as pyridinyl imidazole drugs, have been identified as putative drugs for anti-inflammatory therapies in the CNS [46], the data presented in this paper suggest that inhibiting such molecules (Erk1/2) may represent a pharmaceutical strategy to restrict BBB deterioration, thereby potentially reducing the morbidity associated with neurobrucellosis.

In summary, the results presented here describe a mechanism whereby *B. abortus*-stimulated platelets can induce HBMEC and other endothelial cell activation, promoting neutrophils and monocytes to traverse the BBB. Moreover, this could contribute to increase the infection of glial cells, generating and/or deteriorating neurobrucellosis and the inflammatory response motivated by glial activation (Figure 8).

4. Materials and Methods

4.1. Ethics Statement

Human platelets, monocytes, and neutrophils were isolated from the blood of healthy adult donors in agreement with the guidelines of the Ethical Committee of the Instituto de Medicina Experimental (protocol number: 20160518-M). All adult blood donors provided their informed consent prior to the study.

4.2. Bacteria and Lipoproteins

B. abortus S2308 was cultured in tryptic soy broth supplemented with yeast extract (Merck, Buenos Aires, Argentina). The number of bacteria on stationary-phase cultures was determined by comparing the optical density at 600 nm with a standard curve. All live *Brucella* manipulations were performed in biosafety level 3 facilities located at the Instituto de Investigaciones Biomédicas en Retrovirus y SIDA (INBIRS, Buenos Aires, Argentina). To obtain heat-killed *B. abortus* (HKBA), bacteria were washed five times for 10 min each in sterile phosphate-buffered saline (PBS), heat-killed at 70 °C for 20 min, aliquoted, and stored at −70 °C until used. The total absence of *B. abortus* viability after heat killing was verified by the absence of bacterial growth on tryptic soy agar.

B. abortus lipidated-outer membrane protein 19 (L-Omp19) and unlipidated-Omp19 (U-Omp19) were obtained as described [31]. Both recombinant proteins contained less than 0.25 endotoxin U/µg of protein as assessed by Limulus Amebocyte Lysates (Associates of Cape Cod Inc., Falmouth, MA, USA).

4.3. Cell Lines

HBMECs were isolated from a brain biopsy of an adult female with epilepsy as previously described [47]. These cells were positive for factor VIII–Rag, carbonic anhydrase IV, and Ulex europaeus agglutinin I. They took up fluorescently labeled low-density lipoprotein and expressed g-glutamyl transpeptidase, thus demonstrating their brain endothelial cell properties [47]. HBMECs were subsequently immortalized by transfection with SV40 large T Ag and maintained their morphological and functional characteristics for at least 30 passages [48]. The cells are polarized and exhibit a transendothelial electric resistance (TEER) of at least 100 ohms/cm^2 [49]. Cells (passage < 30) were cultured in tissue culture flasks in Roswell Park Memorial Institute (RPMI) medium 1640 (Life Technologies, Grand Island, NE, USA) supplemented 10% with heat-inactivated fetal bovine serum (FBS) (Life Technologies), 10% NuSerum IV (Becton Dickinson, Bedford, OH, USA), 1% modified Eagle's medium nonessential amino acids (Life Technologies), sodium pyruvate (1 mM), L-glutamine (2 mM), 1% MEM vitamin solution (Life Technologies), penicillin (100 U/mL) and streptomycin (100 µg/mL). Human microvascular endothelial cells (HMEC-1) were obtained from ATCC® (CRL-3243™, Manassas, VA, USA). Cells were grown in Dulbecco's Modified Eagle's (DMEM) medium (Life Technologies) containing 10% FBS (Natocor, Córdoba, Argentina), 10 µg/mL hydrocortisone, 1 ng/mL epidermal growth factor (BD Pharmingen, San Diego, CA, USA), L-glutamine (2 mM), penicillin (100 U/mL), and streptomycin (100 µg/mL). All cell cultures were incubated at 37 °C in a humidified atmosphere of 5% CO_2. Human umbilical vascular endothelial cells (HUVECs) were obtained as described previously [10,50]. Briefly, umbilical vascular tissue was treated with collagenase for digestion. Cells were seeded until confluence on 1% gelatin-coated 25 cm^2 tissue culture flasks and identified by their cobblestone morphology and von Willebrand factor (VWF) antibody (Immunotech, Ocala, FL, USA) binding. Cells were grown in RPMI 1640 medium (Gibco) supplemented with 10% FBS (Life Technologies), heparin (100 µg/mL), endothelial cell growth factor (50 µg/mL), sodium pyruvate (2 mM), L-glutamine (2 mM), penicillin (100 U/mL), and streptomycin (100 µg/mL) at 37 °C in a humidified 5% CO_2 incubator. HUVECs used for experiments were kept between the first and third culture passage.

4.4. Platelet Purification and Stimulation

Platelets were obtained from whole blood from healthy adult human donors as described previously [14]. Briefly, blood samples were collected into tubes containing sodium citrate (Merck) and centrifuged. The platelet-rich plasma was collected and centrifuged in presence of 75 nM prostaglandin I2 (Cayman Chemical, Ann Arbor, MI, USA). Platelets were then washed with RPMI 1640 medium. Finally, platelets were resuspended in RPMI 1640 medium. Platelets were incubated with *B. abortus* (1×10^7/mL) (PLT:B.a. ratio of 1:1) for 24 h in RPMI 1640 medium with 10% FBS (Life Technologies) and L-glutamine (2 mM). In addition, platelets were incubated with HKBA (1×10^8 bacteria/mL), L-Omp19, U-Omp19 (both 500 ng/mL), or thrombin (0.1 U/mL) (Sigma Aldrich, St. Louis, MO, USA). Then, supernatants were collected, sterilized by filtration, ultracentrifuged when mentioned (at 100,000× *g* for 5 h at 4 °C), and stored at −70 °C until they were used.

4.5. Endothelial Cell Treatment

HBMEC, HMEC-1, and HUVEC were cultured in 48 wells plate (5×10^4/0.2 mL). To co-culture infection, platelets were added (cell:platelets ratio, 1:100) and endothelial cells–platelets cultures were infected by *B. abortus* (multiplicity of infection of 100). In all cases, the infection was performed for 2 h in medium containing no antibiotics. Then, cells were maintained for 24 h in the presence of antibiotics (100 μg/mL gentamicin and 50 μg/mL streptomycin) to kill the remaining extracellular bacteria. For experiments with platelet-conditioned media, HBMEC, HUVEC, and HMEC-1 cells were treated with 0.2 mL of diluted supernatants from *B. abortus*-stimulated platelets for 24 h. Culture supernatants from *Brucella*-infected astrocytes and recombinant human IL-1β were used as control. In all cases, cells were harvested to determine cell surface molecule expression by flow cytometry. Supernatants from stimulated endothelial cells were collected and stored at −70 °C until they were used.

4.6. Erk1/2 Signaling Pathway

HBMECs were treated with Erk1/2 MAPK pharmacological inhibitor PD98059 (50 μM) (Calbiochem, San Diego, CA, USA) or vehicle (dimethyl sulfoxide) 2 h before the stimulation with supernatants and the inhibitor were kept throughout the experiment, based on previous report [7].

4.7. Measurement of Cytokine and Chemokine Concentrations

Human IL-6, IL-8, and CCL-2 concentrations were quantified in supernatants harvested from HBMECs and HMEC-1 treated with supernatants from *B. abortus*-stimulated platelets by Sandwich ELISA using paired cytokine-specific mAbs according to the manufacturer's instructions (BD Pharmingen).

4.8. Determination of Cell Surface Molecules by Flow Cytometry

ICAM-1 and CD40 surface expression was determined by flow cytometry. For this, treated HBMECs or HMEC-1 were washed and stained with a PE-labeled antibody (Ab) against human ICAM-1 (CD54) (clone HA58, BD Pharmingen), PE-labeled Ab against human CD40 (clone 5C3; BioLegend, San Diego, CA, USA) or the PE-labeled isotype-matched control Ab (BD Pharmingen). Labeled cells were analyzed on a FACSCalibur flow cytometer (BD Biosciences, San Diego, CA, USA), and data were processed using FlowJo software.

4.9. Neutrophil and Monocytes Transendothelial Migration Assay

Peripheral blood mononuclear cells (PBMCs) and neutrophils were separated by Ficoll-Hypaque (GE Healthcare, Uppsala, Sweden) gradient centrifugation. Human neutrophils were isolated by sedimentation of erythrocytes in 6% dextran and hypotonic lysis as previously described [26]. Monocytes were then purified from PBMCs by Percoll (GE Healthcare) gradient. Both types of cells were resuspended in RPMI 1640 supplemented with 10% FBS. Cell purity was 90% as determined by flow cytometry for both populations. Viability of cells was more than 95% in all the experiments as measured by trypan blue exclusion test.

HBMEC monolayers were established from 20,000 cells per insert on 3-μm pore size membrane Transwell plates of 6.5-mm diameter insert (Corning-Costar, Acton, MA, USA) previously treated with rat tail collagen (50 mg/mL in 1% acetic acid) (BD Biosciences) and neutralized in a saturated atmosphere of ammonium hydroxide. After 5 days, when cellular confluence was reached TEER and passive diffusion of horseradish peroxidase was measured as an indication of monolayer integrity [5]. Then, monolayers were incubated for 24 h with supernatants from *B. abortus*-stimulated platelets. Supernatants from platelets alone as well as non-treated HBMECs were used as negative control. Culture supernatants from *Brucella*-infected astrocytes and recombinant human IL-1β were used as positive control. After that, monolayers were washed and neutrophils or monocytes (1×10^5 cells) were added to the upper chamber in fresh medium. Plates were incubated for 3 h at 37 °C in 5% CO_2 and transmigrated cells to the lower chamber were counted on a hemocytometer.

4.10. Statistical Analysis

Results were analyzed with one-way ANOVA followed by Tukey post-test using the GraphPad Prism 5.0 software.

Author Contributions: A.M.R., A.T., P.B., and G.H.G. conceived and designed the experiments. A.M.R., A.T., A.P.M., M.C.M., and M.V.D. performed the experiments. K.S.K. supported the work with key suggestions and helped with data interpretation. A.M.R. and G.H.G. analyzed the data and wrote the manuscript. G.H.G. supervised experiments, interpreted the data, and supervised the manuscript. All authors reviewed the manuscript. All authors have read and agreed to the published version of the manuscript.

Funding: This research was funded by grants from the Agencia Nacional de Promoción Científica y Tecnológica (ANPCYT-Argentina) [PICT 2014-1111, 2014-1925, 2015-0316, 2016-0356, 2016-1493, 2017-1393, 2017-1905], by grants from CONICET (Argentina) [PIP 0200], by grant UBACYT from University of Buenos Aires [20020170100320BA] (Argentina).

Acknowledgments: We thank Horacio Salomón and the staff of the Instituto de Investigaciones Biomédicas en Retrovirus y SIDA (Universidad de Buenos Aires) for allowing us to use the biosafety level 3 laboratory facilities. A.M.R., M.V.D., P.B., and G.H.G. are members of the Research Career of CONICET (Argentina). A.T. and M.C.M. are recipients of fellowships from Consejo Nacional de Investigaciones Científicas y Técnicas (Argentina).

Conflicts of Interest: The authors declare no conflict of interest.

References

1. Johansson, B.B. The Physiology of the Blood-Brain Barrier. *Neurotransm. Interact. Cogn. Funct.* **1990**, *274*, 25–39. [CrossRef]
2. Almutairi, M.M.A.; Gong, C.; Xu, Y.G.; Chang, Y.; Shi, H. Factors controlling permeability of the blood–brain barrier. *Cell. Mol. Life Sci.* **2015**, *73*, 57–77. [CrossRef] [PubMed]
3. Young, E.J.; Corbel, M.J. *Brucellosis in Latin America. Brucellosis: Clinical and Laboratory Aspects*; CRC Press: Boca Raton, FL, USA, 1989; pp. 151–162.
4. Giambartolomei, G.H.; Wallach, J.C.; Baldi, P.C. Neurobrucellosis. In *Encephalitis: Diagnosis and Treatment*; The Egerton Group: New York, NY, USA, 2008; Volume 14, pp. 255–272.
5. Miraglia, M.C.; Rodriguez, A.M.; Barrionuevo, P.; Rodríguez, J.; Kim, K.S.; Dennis, V.A.; Delpino, M.V.; Giambartolomei, G.H. Brucella abortus Traverses Brain Microvascular Endothelial Cells Using Infected Monocytes as a Trojan Horse. *Front. Microbiol.* **2018**, *8*, 200. [CrossRef] [PubMed]
6. Samartino, C.G.; Delpino, M.V.; Godoy, C.P.; Di Genaro, M.S.; Pasquevich, K.A.; Zwerdling, A.; Barrionuevo, P.; Mathieu, P.; Cassataro, J.; Pitossi, F.; et al. Brucella abortus Induces the Secretion of Proinflammatory Mediators from Glial Cells Leading to Astrocyte Apoptosis. *Am. J. Pathol.* **2010**, *176*, 1323–1338. [CrossRef]

7. Miraglia, M.C.; Scian, R.; Samartino, C.G.; Barrionuevo, P.; Rodriguez, A.M.; E Ibañez, A.; Coria, L.M.; Velásquez, L.N.; Baldi, P.C.; Cassataro, J.; et al. Brucella abortus induces TNF-α-dependent astroglial MMP-9 secretion through mitogen-activated protein kinases. *J. Neuroinflamm.* **2013**, *10*, 47. [CrossRef]
8. Rodríguez, A.M.; Delpino, M.V.; Miraglia, M.C.; Franco, M.M.C.; Barrionuevo, P.; Dennis, V.A.; Oliveira, S.C.; Giambartolomei, G.H. Brucella abortus-activated microglia induce neuronal death through primary phagocytosis. *Glia* **2017**, *65*, 1137–1151. [CrossRef]
9. Miraglia, M.C.; Franco, M.M.C.; Rodriguez, A.M.; Bellozi, P.M.Q.; Ferrari, C.C.; Farias, M.I.; Dennis, V.A.; Barrionuevo, P.; de Oliveira, A.C.P.; Pitossi, F.; et al. Glial Cell-Elicited Activation of Brain Microvasculature in Response to Brucella abortus Infection Requires ASC Inflammasome-Dependent IL-1beta Production. *J. Immunol.* **2016**, *196*, 3794–3805. [CrossRef]
10. Landoni, V.I.; Schierloh, P.; Nebel, M.D.C.; Fernandez, G.C.; Calatayud, C.; Lapponi, M.J.; Isturiz, M.A. Shiga Toxin 1 Induces on Lipopolysaccharide-Treated Astrocytes the Release of Tumor Necrosis Factor-alpha that Alter Brain-Like Endothelium Integrity. *PLoS Pathog.* **2012**, *8*, e1002632. [CrossRef]
11. Abbott, N.J. Astrocyte-endothelial interactions and blood-brain barrier permeability. *J. Anat.* **2002**, *200*, 629–638. [CrossRef]
12. Holinstat, M. Normal platelet function. *Cancer Metastasis Rev.* **2017**, *36*, 195–198. [CrossRef]
13. Carestia, A.; Kaufman, T.; Rivadeneyra, L.; Landoni, V.I.; Pozner, R.G.; Negrotto, S.; D'Atri, L.P.; Gómez, R.M.; Schattner, M. Mediators and molecular pathways involved in the regulation of neutrophil extracellular trap formation mediated by activated platelets. *J. Leukoc. Biol.* **2015**, *99*, 153–162. [CrossRef] [PubMed]
14. Trotta, A.; Velásquez, L.N.; Milillo, M.A.; Delpino, M.V.; Rodriguez, A.M.; Landoni, V.I.; Giambartolomei, G.H.; Pozner, R.G.; Barrionuevo, P. Platelets Promote Brucella abortus Monocyte Invasion by Establishing Complexes with Monocytes. *Front. Immunol.* **2018**, *9*, 1000. [CrossRef] [PubMed]
15. Blair, P.; Flaumenhaft, R. Platelet α-granules: Basic biology and clinical correlates. *Blood Rev.* **2009**, *23*, 177–189. [CrossRef] [PubMed]
16. May, A.E.; Seizer, P.; Gawaz, M. Platelets: Inflammatory Firebugs of Vascular Walls. *Arter. Thromb. Vasc. Biol.* **2008**, *28*, s5–s10. [CrossRef]
17. Fréchette, J.-P.; Martineau, I.; Gagnon, G. Platelet-rich Plasmas: Growth Factor Content and Roles in Wound Healing. *J. Dent. Res.* **2005**, *84*, 434–439. [CrossRef]
18. Cox, D.; McConkey, S.; McConkey, S. The role of platelets in the pathogenesis of cerebral malaria. *Cell. Mol. Life Sci.* **2009**, *67*, 557–568. [CrossRef]
19. Davidson, D.C.; Hirschman, M.P.; Sun, A.; Singh, M.V.; Kasischke, K.; Maggirwar, S.B. Excess Soluble CD40L Contributes to Blood Brain Barrier Permeability In Vivo: Implications for HIV-Associated Neurocognitive Disorders. *PLoS ONE* **2012**, *7*, e51793. [CrossRef]
20. Garciarena, C.D.; McHale, T.M.; Watkin, R.L.; Kerrigan, S.W. Coordinated Molecular Cross-Talk between Staphylococcus aureus, Endothelial Cells and Platelets in Bloodstream Infection. *Pathogens* **2015**, *4*, 869–882. [CrossRef]
21. Trotta, A.; Milillo, M.A.; Serafino, A.; Castillo, L.A.; Weiss, F.B.; Delpino, M.V.; Giambartolomei, G.H.; Fernández, G.C.; Barrionuevo, P. Brucella abortus–infected platelets modulate the activation of neutrophils. *Immunol. Cell Biol.* **2020**. [CrossRef]
22. Lyck, R.; Enzmann, G. The physiological roles of ICAM-1 and ICAM-2 in neutrophil migration into tissues. *Curr. Opin. Hematol.* **2015**, *22*, 53–59. [CrossRef]
23. Ramirez, S.H.; Fan, S.; Dykstra, H.; Reichenbach, N.; Del Valle, L.; Potula, R.; Phipps, R.P.; Maggirwar, S.B.; Persidsky, Y. Dyad of CD40/CD40 ligand fosters neuroinflammation at the blood-brain barrier and is regulated via JNK signaling: Implications for HIV-1 encephalitis. *J. Neurosci.* **2010**, *30*, 9454–9464. [CrossRef] [PubMed]
24. Pollak, C.N.; Delpino, M.V.; Fossati, C.A.; Baldi, P.C. Outer Membrane Vesicles from Brucella abortus Promote Bacterial Internalization by Human Monocytes and Modulate Their Innate Immune Response. *PLoS ONE* **2012**, *7*, e50214. [CrossRef]
25. Giambartolomei, G.H.; Zwerdling, A.; Cassataro, J.; Bruno, L.; Fossati, C.A.; Philipp, M.T. Lipoproteins, not lipopolysaccharide, are the key mediators of the proinflammatory response elicited by heat-killed Brucella abortus. *J. Immunol.* **2004**, *173*, 4635–4642. [CrossRef] [PubMed]
26. Zwerdling, A.; Delpino, M.V.; Pasquevich, K.A.; Barrionuevo, P.; Cassataro, J.; Samartino, C.G.; Giambartolomei, G.H. Brucella abortus activates human neutrophils. *Microbes Infect.* **2009**, *11*, 689–697. [CrossRef] [PubMed]

27. Miao, Z.; Dong, Y.; Fang, W.; Shang, D.; Liu, D.; Zhang, K.; Li, B.; Chen, Y.-H. VEGF Increases Paracellular Permeability in Brain Endothelial Cells via Upregulation of EphA2. *Anat. Rec. Adv. Integr. Anat. Evol. Biol.* **2014**, *297*, 964–972. [CrossRef] [PubMed]
28. Weyrich, A.S.; Zimmerman, G.A. Platelets: Signaling cells in the immune continuum. *Trends Immunol.* **2004**, *25*, 489–495. [CrossRef] [PubMed]
29. Iwasaki, A.; Medzhitov, R. Regulation of Adaptive Immunity by the Innate Immune System. *Science* **2010**, *327*, 291–295. [CrossRef]
30. Gawaz, M.; Langer, H.; May, A.E. Platelets in inflammation and atherogenesis. *J. Clin. Investig.* **2005**, *115*, 3378–3384. [CrossRef]
31. Semple, J.W.; Italiano, J.E.; Freedman, J. Platelets and the immune continuum. *Nat. Rev. Immunol.* **2011**, *11*, 264–274. [CrossRef]
32. Danese, S.; Dejana, E.; Fiocchi, C. Immune regulation by microvascular endothelial cells: Directing innate and adaptive immunity, coagulation, and inflammation. *J. Immunol.* **2007**, *178*, 6017–6022. [CrossRef]
33. Pober, J.S.; Sessa, W.C. Evolving functions of endothelial cells in inflammation. *Nat. Rev. Immunol.* **2007**, *7*, 803–815. [CrossRef] [PubMed]
34. Singh, M.V.; Davidson, D.C.; Jackson, J.W.; Singh, V.B.; Silva, J.; Ramirez, S.H.; Maggirwar, S.B. Characterization of platelet-monocyte complexes in HIV-1-infected individuals: Possible role in HIV-associated neuroinflammation. *J. Immunol.* **2014**, *192*, 4674–4684. [CrossRef] [PubMed]
35. Cox, D.; Kerrigan, S.W.; Watson, S. Platelets and the innate immune system: Mechanisms of bacterial-induced platelet activation. *J. Thromb. Haemost.* **2011**, *9*, 1097–1107. [CrossRef] [PubMed]
36. Fitzgerald, J.R.; Foster, T.J.; Cox, D. The interaction of bacterial pathogens with platelets. *Nat. Rev. Genet.* **2006**, *4*, 445–457. [CrossRef] [PubMed]
37. Gawaz, M.; Neumann, F.-J.; Dickfeld, T.; Koch, W.; Laugwitz, K.-L.; Adelsberger, H.; Langenbrink, K.; Page, S.; Neumeier, D.; Schoömig, A.; et al. Activated Platelets Induce Monocyte Chemotactic Protein-1 Secretion and Surface Expression of Intercellular Adhesion Molecule-1 on Endothelial Cells. *Circulation* **1998**, *98*, 1164–1171. [CrossRef]
38. Ley, K.; Reutershan, J. Leucocyte-Endothelial Interactions in Health and Disease. *Handb. Exp. Pharmacol.* **2006**, *176*, 97–133. [CrossRef]
39. Vieira-De-Abreu, A.; Campbell, R.A.; Weyrich, A.S.; Zimmerman, G.A. Platelets: Versatile effector cells in hemostasis, inflammation, and the immune continuum. *Semin. Immunopathol.* **2011**, *34*, 5–30. [CrossRef]
40. Henn, V.; Slupsky, J.R.; Gräfe, M.; Anagnostopoulos, I.; Foörster, R.; Müller-Berghaus, G.; Kroczek, R.A. CD40 ligand on activated platelets triggers an inflammatory reaction of endothelial cells. *Nature* **1998**, *391*, 591–594. [CrossRef]
41. Zuchtriegel, G.; Uhl, B.; Puhr-Westerheide, D.; Pörnbacher, M.; Lauber, K.; Krombach, F.; Reichel, C.A. Platelets Guide Leukocytes to Their Sites of Extravasation. *PLoS Biol.* **2016**, *14*, e1002459. [CrossRef]
42. Halling, S.M.; Peterson-Burch, B.D.; Bricker, B.J.; Zuerner, R.L.; Qing, Z.; Li, L.-L.; Kapur, V.; Alt, D.P.; Olsen, S.C. Completion of the Genome Sequence of Brucella abortus and Comparison to the Highly Similar Genomes of Brucella melitensis and Brucella suis. *J. Bacteriol.* **2005**, *187*, 2715–2726. [CrossRef]
43. Tibor, A.; Decelle, B.; Letesson, J.-J. Outer Membrane Proteins Omp10, Omp16, and Omp19 of Brucella spp. Are Lipoproteins. *Infect. Immun.* **1999**, *67*, 4960–4962. [CrossRef] [PubMed]
44. Sun, Y.; Liu, W.-Z.; Liu, T.; Feng, X.; Yang, N.; Zhou, H.-F. Signaling pathway of MAPK/ERK in cell proliferation, differentiation, migration, senescence and apoptosis. *J. Recept. Signal Transduct.* **2015**, *35*, 600–604. [CrossRef] [PubMed]
45. Aramoto, H.; Breslin, J.W.; Pappas, P.J.; Hobson, R.W.; Durán, W.N. Vascular endothelial growth factor stimulates differential signaling pathways in in vivo microcirculation. *Am. J. Physiol. Circ. Physiol.* **2004**, *287*, H1590–H1598. [CrossRef] [PubMed]
46. Barone, F.C.; Irving, E.; Ray, A.; Lee, J.; Kassis, S.; Kumar, S.; Badger, A.; Legos, J.J.; A Erhardt, J.; Ohlstein, E.; et al. Inhibition of p38 mitogen-activated protein kinase provides neuroprotection in cerebral focal ischemia. *Med. Res. Rev.* **2001**, *21*, 129–145. [CrossRef]
47. Stins, M.F.; Gilles, F.; Kim, K.S. Selective expression of adhesion molecules on human brain microvascular endothelial cells. *J. Neuroimmunol.* **1997**, *76*, 81–90. [CrossRef]
48. Stins, M.F.; Badgera, J.; Kim, K.S. Bacterial invasion and transcytosis in transfected human brain microvascular endothelial cells. *Microb. Pathog.* **2001**, *30*, 19–28. [CrossRef]

49. Nizet, V.; Kim, K.S.; Stins, M.; Jonas, M.; Chi, E.Y.; Nguyen, D.; Rubens, C.E. Invasion of brain microvascular endothelial cells by group B streptococci. *Infect. Immun.* **1997**, *65*, 5074–5081. [CrossRef]
50. Ferrero, M.C.; Bregante, J.; Delpino, M.V.; Barrionuevo, P.; A Fossati, C.; Giambartolomei, G.H.; Baldi, P.C. Proinflammatory response of human endothelial cells to Brucella infection. *Microbes Infect.* **2011**, *13*, 852–861. [CrossRef]

pathogens

MDPI

Review

Adhesins of *Brucella*: Their Roles in the Interaction with the Host

Magalí G. Bialer [1], Gabriela Sycz [1,†], Florencia Muñoz González [2,3], Mariana C. Ferrero [2,3], Pablo C. Baldi [2,3,*] and Angeles Zorreguieta [1,4,*]

1 Fundación Instituto Leloir (FIL), IIBBA (CONICET-FIL), Buenos Aires 1405, Argentina; mbialer@leloir.org.ar (M.G.B.); gabrielasycz@gmail.com (G.S.)
2 Cátedra de Inmunología, Facultad de Farmacia y Bioquímica, Universidad de Buenos Aires, Buenos Aires 1113, Argentina; fmgonzalez@ffyb.uba.ar (F.M.G.); ferrerom@ffyb.uba.ar (M.C.F.)
3 Instituto de Estudios de la Inmunidad Humoral (IDEHU), CONICET-Universidad de Buenos Aires, Buenos Aires 1113, Argentina
4 Departamento de Química Biológica, Facultad de Ciencias Exactas y Naturales, Universidad de Buenos Aires, Buenos Aires 1428, Argentina
* Correspondence: pablobal@ffyb.uba.ar (P.C.B.); azorreguieta@leloir.org.ar (A.Z.)
† Current address: Department of Molecular Microbiology, Washington University School of Medicine in St. Louis, 660 S Euclid Ave (63110), St. Louis, MO 63110, USA.

Received: 30 September 2020; Accepted: 5 November 2020; Published: 12 November 2020

Abstract: A central aspect of *Brucella* pathogenicity is its ability to invade, survive, and replicate in diverse phagocytic and non-phagocytic cell types, leading to chronic infections and chronic inflammatory phenomena. Adhesion to the target cell is a critical first step in the invasion process. Several *Brucella* adhesins have been shown to mediate adhesion to cells, extracellular matrix components (ECM), or both. These include the sialic acid-binding proteins SP29 and SP41 (binding to erythrocytes and epithelial cells, respectively), the BigA and BigB proteins that contain an Ig-like domain (binding to cell adhesion molecules in epithelial cells), the monomeric autotransporters BmaA, BmaB, and BmaC (binding to ECM components, epithelial cells, osteoblasts, synoviocytes, and trophoblasts), the trimeric autotransporters BtaE and BtaF (binding to ECM components and epithelial cells) and Bp26 (binding to ECM components). An in vivo role has also been shown for the trimeric autotransporters, as deletion mutants display decreased colonization after oral and/or respiratory infection in mice, and it has also been suggested for BigA and BigB. Several adhesins have shown unipolar localization, suggesting that *Brucella* would express an adhesive pole. Adhesin-based vaccines may be useful to prevent brucellosis, as intranasal immunization in mice with BtaF conferred high levels of protection against oral challenge with *B. suis*.

Keywords: *Brucella*; adhesins; Ig-like domain; monomeric autotransporters; trimeric autotransporters; extracellular matrix; polar localization; virulence factors; vaccine candidates; fibronectin

---

1. Introduction

Brucella spp. are Gram-negative bacteria that infect several animal species and can be transmitted to humans by several routes, producing one of the most common zoonotic diseases worldwide. A central aspect of *Brucella* pathogenicity is its ability to invade, survive, and replicate in several phagocytic and non-phagocytic cell types, leading not only to chronic infections but also to chronic inflammatory phenomena in different tissues. For both phagocytic and non-phagocytic cells, the first step of the invasion process involves interactions between surface molecular factors of *Brucella* and the host cell, leading to the cellular adhesion of the pathogen. Several *Brucella* proteins have been shown to be involved in the adhesion of this bacterium to different cell types and/or to extracellular matrix

(ECM) components. In this review, we describe the main characteristics of these *Brucella* adhesins in the broader context of bacterial adhesins, and how they contribute to the cellular infectious process of brucellae.

2. *Brucella* Infection and Clinical Manifestations

With more than 500,000 new cases annually, human brucellosis continues to be one of the commonest zoonotic diseases worldwide. Although this disease has been eradicated in some developed countries, it still constitutes a public health problem in Latin America, the Middle East, North and East Africa, and South and Central Asia [1]. Moreover, the disease is still present in some European countries.

Brucella spp. are Gram-negative non-capsulated and non-sporulated bacilli or cocobacilli that lack cilia or flagella. Despite the great number of *Brucella* species identified, which may infect domestic animals and wild animals, only *B. melitensis* (goats and sheep), *B. suis* (pigs), *B. abortus* (cattle), and, to a minor extent, *B. canis* (dogs) are linked to human brucellosis. Different *Brucella* species yield smooth (S) or rough (R) colonies when cultured in agar, which is a difference that is directly related to the structure of the lipopolysaccharide (LPS). The LPS is divided into three regions: the lipid A (the innermost portion), a polysaccharidic core, and the O polysaccharide (the outermost portion). Smooth species (*B. melitensis*, *B. suis*, *B. abortus*, and others) produce a "complete" LPS (S-LPS) containing the three portions, whereas rough species (*B. canis* and *B. ovis*) produce a rough LPS (R-LPS) that lacks the O polysaccharide.

Brucella infection in humans is mainly acquired through the consumption of raw animal products, inhalation of contaminated aerosols in slaughterhouses, rural settings or laboratories [2], and contact of the abraded skin with contaminated tissues or materials. Less frequently, accidental infection with attenuated vaccine strains [3–6] and vertical transmission have been reported [7].

Human brucellosis has a wide spectrum of clinical manifestations, which depend on the stage of the disease and the organs and systems involved. The disease usually presents as a febrile illness accompanied by myalgia, arthralgia, and hepatomegaly, and may evolve with an uncomplicated course or may present complications involving particular organs or systems [8]. Osteoarticular involvement is the most common focal complication [9]. The diversity of tissues that can be affected by *Brucella* is likely related to its ability to invade, survive, and replicate in several phagocytic and non-phagocytic cell types, as explained below. Despite its tendency to produce chronic illness and even disabling disease, human brucellosis is only rarely fatal. In animals, the most prevalent manifestations are abortions, reduced fertility, weight loss, and reduced milk production [10].

3. *Brucella* Entry into Host Cells

A central aspect of *Brucella* pathogenicity is its ability to invade, survive, and replicate in several cell types, leading not only to chronic infections but also to chronic inflammatory phenomena that explain most of the clinical manifestations of brucellosis [11]. Most initial research was performed using murine macrophagic cell lines [12], bovine macrophages [13], human monocytes [14], and widely used non-phagocytic cell lines such as HeLa (human cervical cells) or Vero (kidney, African green monkey) [15,16]. However, further in vitro studies revealed that *Brucella* is capable to infect and replicate in human osteoblasts [17], synoviocytes [18], trophoblasts [19], endothelial cells [20], lung epithelial cells [21], dendritic cells [22], and hepatocytes [23] as well as in murine alveolar macrophages [24], canine trophoblasts and phagocytes [25], and ovine testis cells lines [26].

The internalization of *Brucella* into the different cell types is a complex multi-stage process. Whatever the host cell (phagocytic or non-phagocytic) and the *Brucella* strain involved (smooth or rough), the first step in this process involves interactions between surface molecular factors of both the host cell and the pathogen leading to cellular binding of the bacterium. In fact, it was shown by scanning electron microscopy that *B. abortus* adheres (as early as 1 h after infection) and forms bacterial aggregates on the surface of host cells in a time-dependent manner [27]. Whereas several of the surface molecular factors involved have been identified, the full repertoire of molecular components and mechanisms acting on either active bacterial penetration or passive uptake of *Brucella* spp. are not fully characterized [28]. For non-opsonized bacteria, internalization into macrophages seems to depend on lipid rafts present in the plasma membrane of these cells [29,30]. It has been shown that lipid rafts-associated molecules, including cholesterol and the ganglioside GM1, are involved in the entry of *B. suis* into murine macrophages under non-opsonic conditions [31]. In addition, a class A scavenger receptor (SR-A) seems to be required for *B. abortus* internalization into macrophages through a lipid raft-mediated mechanism [32]. These three host molecules have been involved in the ability of naturally rough *Brucella* species (*B. ovis*, *B. canis*) to infect murine macrophages [33]. Although these lipid raft-associated molecules have a role in *Brucella* internalization in macrophages, it has not been determined whether they participate in bacterial adhesion or, alternatively, only contribute to bacterial penetration or uptake.

In addition to these lipid raft-associated molecules, other host components have been identified as being involved in the interaction between brucellae and the host cells. Sialic acid-containing molecules were proposed to be involved in the interaction of brucellae with macrophages and epithelial cells [27]. GM1 is a sialylated molecule, which may perhaps explain its role in lipid raft-mediated internalization in macrophages. This study also produced evidence suggesting that cell surface heparan sulphate molecules may be involved in *Brucella* binding to epithelial cells. Based on the hypothesis that the interactions of *Brucella* with the ECM contribute to the spread of the bacteria through tissue barriers, the ability of the pathogen to bind to ECM constituents was also explored. It was shown that *B. abortus* binds in a dose-dependent manner to immobilized fibronectin and vitronectin and, to a lesser extent, to chondroitin sulphate, collagen, and laminin [27].

As mentioned above, adhesion to host cells is the first step of the infectious cycle of many pathogens. Most bacterial pathogens express adhesins and other molecules that mediate the binding to a wide range of cell surface molecules and ECM components depending on the lifestyle of the microorganism. The fact that *Brucella* species can bind to the cell surface and ECM components strongly suggests the expression of bacterial molecules involved in such an interaction. Although not formally shown to be involved in adhesion, *Brucella* LPS has been linked to the internalization of the pathogen in macrophages. Smooth *B. abortus* strains expressing a complete LPS (including the O-polysaccharide) enter macrophages through lipid-rafts, whereas a rough mutant does not [30,34]. However, naturally rough *Brucella* species (*B. ovis, B. canis*) seem to use lipid rafts for entry [33], suggesting that lipid raft-mediated internalization of brucellae does not depend on O-polysaccharide expression. A role for some outer membrane proteins, namely Omp22 and Omp25, in *Brucella* binding or internalization has also been suggested. Targeted inactivation of their corresponding genes impaired internalization of rough *B. ovis* but not that of *B. abortus* [35,36]. Moreover, *B. abortus* mutants were more adherent than the wild-type strain. While the role of the LPS and outer membrane proteins in the ability of *Brucella* to adhere to cells or ECM requires further clarification, more recent studies have led to the identification of bacterial adhesins clearly involved in these adhesion processes (see below).

Upon entry into the host cells, *Brucella* organisms initiate an intracellular cycle that involves a sequential traffic through the endocytic, secretory, and autophagic compartments. Bacterial effectors delivered inside the infected cells through a type IV secretion system encoded by the *virB* operon are essential to accomplish these steps [15,37–39]. The O polysaccharide of the LPS is also involved in the ability of *Brucella* to establish intracellular infections. Phagosomes containing smooth strains of *B. suis* do not fuse with lysosomes, at least in murine macrophages, whereas those harboring rough mutants rapidly fuse [40]. This seems to be related to the fact that only the naturally smooth strains enter the cells through lipid-rafts and can inhibit phagosome-lysosome fusion [30,34].

4. Bacterial Adhesins

Most pathogenic bacteria interact with their hosts through adhesive molecules (adhesins) that are exposed on their cell surfaces. Since adhesion to host cells can also stimulate immune activation, several bacteria produce a surface layer (i.e., capsular polysaccharide) that prevents immune recognition or phagocytosis. For this reason, they often express adhesins on polymeric structures that extend out from the cell surface at a prudential distance. For some bacteria, attachment to the host cell surface is also crucial for effector injection through complex secretion systems. Lastly, adhesion to the host cell is the previous step to internalization for those bacteria whose strategy to achieve proliferation and survival is the intracellular life [41,42]. The adhesins can be grouped into two types: (1) filamentous (fimbrial) adhesins consisting of complex structures made up of multiple subunits and (2) non-fimbrial adhesins that can be monomeric or trimeric proteins.

Fimbrial adhesins are a varied group of polymeric fibers that are visible using electron microscopy. In Gram-negative bacteria, these adhesins can be classified into: (1) the chaperone-usher pili (CUP), (2) the alternative chaperone-usher pathway pili, (3) Type IV pili, and (4) pili assembled by the extracellular nucleation-precipitation pathway (curli) [41,43]. The subunit at the tip of the CUP pili is a lectin that can bind sugar-containing molecules on the host cell surfaces [44]. The Type IV pili are long filaments composed of pilin subunits assembled into bundles, which are involved in diverse functions including bacterial twitching motility, auto-aggregation, and attachment to host cells [45]. Curli are involved in many physiological and pathogenic processes such as biofilm formation and host cell adhesion and invasion. The curli are assembled via the nucleation-precipitation pathway and display structural similarities with functional amyloids [46,47]. As mentioned below, *Brucella* spp. do not seem to express fimbrial adhesins.

Non-fimbrial adhesins include adhesins that belong to the RTX (repeat in toxin) protein family and those that correspond to type V secretion systems (T5SS), which are also called autotransporter proteins.

RTX adhesins are secreted by a type 1 secretion system (T1SS) that has three components: an inner-membrane ABC (ATP binding cassette) transporter, a membrane fusion protein, and an outer-membrane pore from the TolC family. The substrates of T1SSs do not harbor an N-terminal cleavable signal peptide but share a structural C-terminal domain that is not cleaved off during the secretion process [48]. The RTX adhesins are usually loosely attached to the bacterial surface and have been implicated in bacteria-to-bacteria interactions during biofilm formation and adhesion to epithelial cells [49].

The T5SSs (subfamilies Va–Ve) play important roles in the interaction of several pathogens with their hosts [50]. Originally, the term "autotransporter" was proposed because it was thought that all the information for its translocation from the inner membrane to the extracellular medium was mostly contained in the protein itself. This concept has changed since other factors, such as chaperones and the BAM (β-barrel Assembly Machinery) system are required for secretion of these proteins. Furthermore, more recently, it was shown that another system, the TAM (Translocation and Assembly Module) complex, is also required for the correct translocation of autotransporters into the outer membrane [51–53]. It was proposed that this complex spanning the periplasmic space might solve the energy problem to translocate proteins through the outer membrane [52]. Therefore, the current model proposes that the TAM and BAM systems would act in a concerted manner [51,54].

The T5SS or autotransporter proteins share common structural and functional characteristics: (1) an N-terminal Sec-dependent signal peptide that mediates the transport from the cytoplasm to the periplasm, (2) a passenger (and functional) domain, and (3) a C-terminal β-barrel domain that forms a pore in the outer membrane through which the passenger domain is translocated to the cell surface [55]. In the subclass Va, the autotransporters are monomeric and the passenger and secretion domains are integrated into the same protein, the β-barrel domain forms a pore of 12 antiparallel β-strands, and the passenger regions consist of highly variable repetitive amino acid motifs. Some of these autotransporters are important virulence factors, playing diverse functions in the interaction with the host. The passenger domains have often enzymatic activity and usually adopt a repetitive β-helix fold extending away from the bacterial cell surface, as demonstrated by the crystal structure of the Pertactin passenger domain [56]. Passengers with enzymatic activity are cleaved off from the surface while adhesion passengers can be retained on the cell surface without cleavage (for a comprehensive review, see Reference [50]). Some adhesins of the monomeric autotransporter family have been described in *Brucella* spp., as explained below.

In the two-partner secretion systems (T5SS type Vb), the passenger and β-translocator domains are encoded by two different genes. Filamentous haemagglutinin adhesins are exported by this type of system. These adhesins are often involved in a tight interaction with a host cell receptor and also in biofilm formation [50].

All members of the trimeric autotransporter Type Vc group that have been characterized so far are implicated in adhesion functions. They usually bind to host receptors or to host ECM components. As detailed in the next section, *Brucella* spp. express adhesins that belong to this subclass of autotransporters. While the overall organization of these proteins is similar to that of the monomeric autotransporters, they contain a shorter C-terminal translocation domain of 50–100 amino acids and the 12 β-strand pore is achieved by protein trimerization. Usually, the passenger domain harbors conserved structural elements named as head, connector, and stalk domains. The combinations of these repeats result in either "lollipop structures" like YadA or as "beads-on-a-string" like BadA [52]. Although the head domains typically mediate adhesion to host targets, the stalk domains can also participate in adhesion functions. Internal regions may serve to extend the head domain away from the bacterial cell surface. Unlike several monomeric autotransporters, trimeric autotransporters are not released into the extracellular space [50].

The Type Ve of T5SS harbors a 12-stranded β-barrel domain and a secreted, monomeric passenger domain that remains attached after translocation. The main difference with type Va autotransporters is that the type Ve have an inverted domain order with the β-barrel at the N-terminal end and the passenger domain at the C-terminus, and, thus, are named as "inverse autotransporters" [57]. Well-known examples are the intimin and invasin from pathogenic *Escherichia coli* and *Yersinia* spp., respectively. The passenger domains of this type of T5SS contain domains with Immunoglobulin (Ig)-like or lectin-like structures. The intimin of enteropathogenic and enterohemorrhagic *E. coli* strains mediates an intimate contact with the Tir receptor, which is delivered by the bacterium to the surface of the host cell. The invasin of *Yersinia* spp. binds directly to β1-integrins on the apical side of gut epithelial cells, which promotes bacterial internalization via endocytosis [58,59].

5. Adhesins of *Brucella*

The genomes of *Brucella* spp. do not harbor loci associated with components of pili or curli that could function as fimbrial adhesins. Furthermore, by electron microscopy, no pilus-like structures have been observed. However, several non-fimbrial adhesins have been identified that were shown to have a role in the interaction with the host. A diagram of these adhesins, showing their domains, is depicted in Figure 1, and additional information is presented in Table 1.

Table 1. Adhesins described in *Brucella* spp.

Adhesin	Organism	KEGG Entry	NCBI Protein ID	Protein Class	Host Ligands Detected	Cellular Adhesion Role	In Vivo Infection Role	Reference
SP29	*B. abortus* 9-941	BruAb2_0373	WP_002965789.1	D-ribose ABC transporter substrate-binding protein	Sialic acid-containing proteins	Erythrocytes	ND	[60]
SP41	*B. abortus* 9-941	BruAb2_0571	WP_002965982.1	ATP-binding cassette transporter	Sialic acid-containing proteins	Epithelial (HeLa)	No role detected in *B. ovis* infections	[61,62]
BigA	*B. abortus* 2308	BAB1_2009	EEP62646.1	Ig-like domain- containing protein	Cell adhesion molecules	Epithelial (HeLa, Caco.2, MDCK)	Potential role in oral infections *	[63,64]
BigB	*B. abortus* 2308	BAB1_2012	WP_002967016.1	Ig-like domain- containing protein	Cell adhesion molecules	Epithelial (HeLa)	Potential role in oral infections *	[63,65]
Bp26	*B. melitensis 16M*	BMEI0536	WP_002964581.1	Uncharacterized	Type I collagen, vitronectin, fibronectin	ND	ND	[66]
BmaC	*B. suis* 1330	BRA1148	WP_006191504.1	Monomeric autotransporter	Fibronectin, type I collagen	Epithelial (HeLa, A549). Synoviocytes. Osteoblasts	ND	[67,68]
BmaA	*B. suis* 1330	BR0173	AAN33380.1	Monomeric autotransporter	Fibronectin, type I collagen	Epithelial (HT 29, Caco.2). Synoviocytes. Osteoblasts. Trophoblasts	ND	[68]
BmaB	*B. suis* 1330	BR2013	AAN30903.1	Monomeric autotransporter	Fibronectin	Synoviocytes. Osteoblasts. Trophoblasts	ND	[68]
BtaE	*B. suis* 1330	BR0072	WP_006191142.1	Trimeric autotransporter	Fibronectin, hyaluronic acid	Epithelial (HeLa, A549)	Mutants display decreased colonization after oral infection	[69]
BtaF	*B. suis* 1330	BR1846	A0A0H3G4K1.1	Trimeric autotransporter	Fibronectin, hyaluronic acid, fetuin, type I collagen	Epithelial (HeLa, A549)	Mutants display decreased colonization after oral or respiratory infection	[70,71]

ND: not determined. (*) A mutant lacking the Bab1-2009-2012 genomic island is attenuated in oral infections in mice.

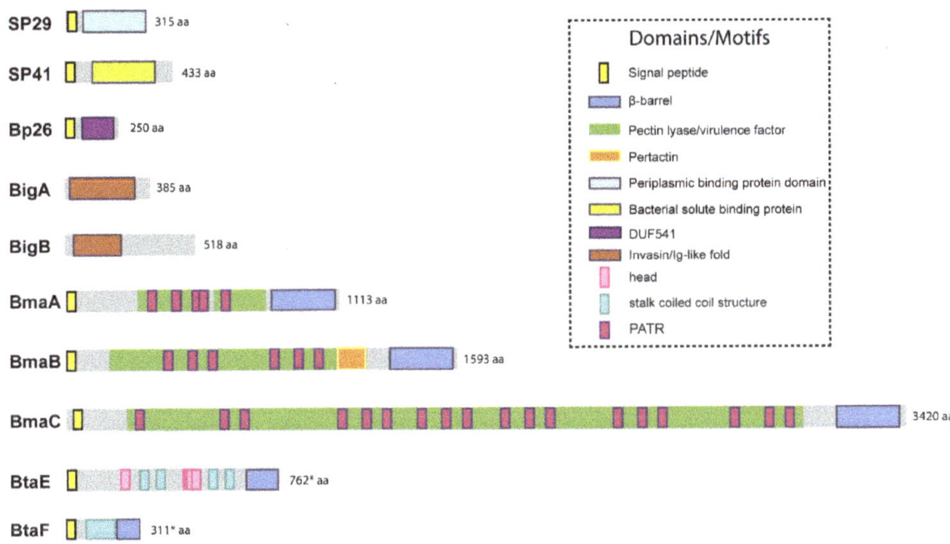

Figure 1. Domain organization of described *Brucella* adhesins. Schematic representation of the adhesins described in *Brucella* spp, showing functional and structural domains predicted by bioinformatics (SignalP 5.0, Pfam, BLAST, InterProScan). Asterisks indicate the cases for which start codons upstream of the annotated ORF were identified and the translation product of the new ORF contained an N-terminal signal peptide with a reliable score. aa: amino acids.

5.1. Unclassified Adhesins

Due to the abundance of carbohydrates in the surface of red cells, hemagglutination tests have been used for the detection and characterization of many lectin-like adhesins in bacterial pathogens. Using this approach, Rocha-Gracia et al. found that *B. abortus* and *B. melitensis* can agglutinate human (A+ and B+), hamster, and rabbit erythrocytes, and that this activity was associated with a bacterial 29-kDa surface protein (SP29) that binds to these cells [60]. Purified SP29 bound directly to rabbit erythrocytes, and this binding was abolished by neuraminidase treatment of red cells, indicating that SP29 binds to sialic acid-containing receptors. The analysis of an internal fragment obtained by peptic digestion suggested that SP29 is a D-ribose-binding periplasmic protein precursor found in *B. melitensis* (BruAb2_0373) (Figure 1). No further characterization of this protein or its importance for *Brucella* pathogenesis has been reported despite the demonstration that *B. melitensis* is able to invade erythrocytes in vivo at least in the mouse model [72]. This later study revealed that *B. melitensis* can adhere to murine erythrocytes as early as 3-h post-infection but is later found mainly in the cytoplasm of these cells. Moreover, erythrocytes represented the major fraction of infected cells in the bloodstream. Purified erythrocytes from infected mice were able to transmit *B. melitensis* infection to naïve mice.

To our best knowledge, the first *Brucella* adhesin for which a functional role was fully characterized in vitro was SP41 (Figure 1, Table 1) [61]. This protein is the predicted product of the *ugpB locus*, which encodes a protein of 433 amino acids with similarity to a periplasmic glycerol-3-phosphate-binding ATP-binding cassette (ABC) transporter protein found in several bacterial species, and harbors a bacterial solute-binding protein domain. Immunofluorescence studies indicated that SP41 is surface exposed, and antibodies directed to SP41 inhibited *B. suis* adherence to HeLa cells. Notably, a Δ*ugpB B. suis* mutant exhibited a significant reduction in the adherence to epithelial cells, supporting the contention that SP41 is an adhesin. Treatment of HeLa cells with neuraminidase abolished SP41 binding to these cells, suggesting the involvement of sialic acid residues in this interaction (Figure 2, Table 1). In contrast, a further study in *B. ovis* did not reveal an effect of *ugpB* deletion on early internalization or intracellular survival of this rough species in murine macrophages (J774.A1 cell line) or HeLa cells [62]. In addition, the deletion had no effect on the ability of *B. ovis* to colonize the spleen after intraperitoneal inoculation in mice. The *ugpB* gene seems to be functional in *B. ovis* as revealed by RT-PCR assays, and the encoded protein differs only by five amino acids from that of *B. suis*. It was argued that other adhesins would be more exposed on the bacterial surface of *B. ovis* due to the absence of O-polysaccharide chains, favoring their interaction with the host cell.

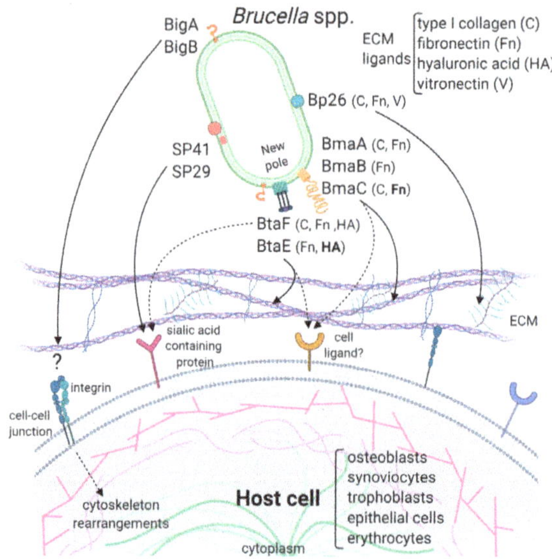

Figure 2. Model of the interaction of *Brucella* adhesins with host cells. The described *Brucella* spp. adhesins are depicted on the bacterial surface and their interactions with ECM components and ligands on the host cell are represented by black arrows. ECM components and cell types to which these adhesins bind are mentioned. ECM ligands in bold are supported by consistent evidence while those in a normal font are supported by indirect evidence. The Bma and Bta proteins are mostly localized at the new pole generated after asymmetric bacterial division. Bipolar localization was shown for BigA, but BigB polarity has not been determined. SP29 is predicted to be periplasmic while SP41 was shown to be exposed on the bacterial surface. It is not clear if Bp26 is localized to the outer membrane or in the periplasm. The cell ligands for Big and Bp26 adhesins have not yet been identified. In addition to ECM, Bma and Bta adhesins could interact with cell surface ligands. Dotted arrows represent putative interactions. Importantly, while all the *Brucella* adhesins characterized to date are shown, they may be not simultaneously expressed on bacteria.

The potential role of the *Brucella* Bp26 protein as an adhesin was recently tested in vitro [66]. The rationale exposed by the authors for testing this protein was not related to structural or homology criteria, but to the fact that Bp26 induces strong antibody responses in infected individuals [73]. Bp26 is a 250 amino acid-predicted protein with a domain of unknown function (DUF541) (Figure 1). The binding properties of Bp26 to ECM components such as type I collagen, fibronectin, vitronectin, and laminin were tested by ELISA and biolayer interferometry. According to the results of these assays, Bp26 binds to both immobilized and soluble type I collagen and vitronectin, and to soluble (but not immobilized) fibronectin, but does not bind to laminin (Figure 2, Table 1). The relevance of Bp26 for in vitro adhesion of *Brucella* to cells or for the outcome of in vivo infections has not been tested.

5.2. Adhesins Containing Ig-Like Domains

A study by Czibener and Ugalde allowed the identification of a pathogenicity island in *B. abortus* (BAB1_2009-2012) whose deletion resulted in a reduced attachment of the bacterium to HeLa cells [63]. Furthermore, the deletion mutant also displayed a reduced capacity to colonize the spleen of mice after oral infection as compared to the wild-type strain. In particular, BAB1_2009 was found to encode a protein that harbors a bacterial Ig-like (BIg-like) domain present in adhesins from the invasin/intimin family [74] (Figure 1). In the following study, the role of this protein (named as BigA) in adhesion to epithelial cells was demonstrated [64]. This study also revealed that BigA is an exposed outer membrane protein and that incubation of the bacteria with antibodies against the Ig-like domain of BigA before infection of HeLa cells reduces the number of intracellular bacteria. While a deletion mutant strain displayed a significant defect in both adhesion and invasion to polarized epithelial cell lines such as Caco-2 (human colon) and Madin–Darby canine kidney (MDCK), overexpression of the *bigA* gene greatly increased them (Table 1). Confocal microscopy analyses showed that the BigA adhesin targets the bacteria to the cell-cell junction membrane in confluent epithelial cells and also induces cytoskeleton rearrangements (Figure 2). A recent publication by the same research group showed that other Ig-like (Blg-like) domain-containing protein (BAB1_2012, named BigB) (Figure 1), encoded by the same locus (BAB1_2009-2012), is also involved in adhesion to epithelial cells and targets proteins involved in cell-cell and cell-matrix interactions (Figure 2) [65]. The Δ*bigB* mutant showed a significant reduction in intracellular bacteria at the early stages of infection both in HeLa cells and in polarized MDCK cells (Table 1). It was further demonstrated by counting fluorescent bacteria that the phenotype on HeLa cells was due to a defect in adhesion. Similar to BigA, recombinant BigB induced profound cytoskeleton rearrangements in HeLa cells (Figure 2). HeLa cells transfected with focal adhesion markers showed changes in focal adhesion sites. It was proposed that, similar to BigA, BigB targets proteins in cell-cell junctions, which, in turn, triggers changes in the cytoskeleton (Figure 2). This work also showed that the BAB1_2011 gene encodes a periplasmic protein (PalA), which is necessary for the proper display of both the BigA and BigB adhesins, indicating that the genomic island is dedicated to the adhesion of *Brucella* to host cells. Although the phenotypes of the *big* mutants have not been tested in vivo, the previous result obtained with the BAB1_2009-2012 deletion mutant strongly suggests that the Big adhesins have a role in vivo.

5.3. Autotransporters

Numerous virulence factors of bacterial pathogens contain domains or motifs related to adhesion to biotic or abiotic surfaces. A comprehensive search for conserved adhesion-associated domains/motifs in the *B. suis* 1330 genome and subsequent phylogenetic analyses revealed the presence of three clearly separated groups of adhesins: (1) monomeric autotransporters (BRA0173, BR2013, BRA1148), (2) trimeric autotransporters (BR0072 and BR1846), and (3) Ig-like domain containing-proteins (BR2009 and BR2012) [75]. Other proteins with no clear associated functions were also identified but not described in this work. Group 2 also included the protein BR0049. We have recently shown that BR0049 is certainly not an adhesin, but is required for the correct insertion of proteins from the autotransporter families (see below) [53]. Group 3 comprised orthologous proteins of the *B. abortus* BigA and BigB proteins described above.

5.3.1. Monomeric Autotransporters

As mentioned above, *B. abortus* binds in a dose-dependent manner to components of the ECM, such as fibronectin and vitronectin [27]. In an attempt to identify *B. suis* genes encoding proteins that might be involved in the binding of brucellae to fibronectin, Posadas et al. [67] panned an M13 phage display library of the *B. suis* 1330 genome against immobilized fibronectin. Several recombinant phages showed affinity to immobilized fibronectin. However, only one expressed a portion of a protein that was predicted to be exposed on the cell surface. This protein corresponded to one of the monomeric autotransporters described above (BRA1148) and was named as BmaC for *Bucella* monomeric autotransporter. This protein exhibits all the characteristics of this protein family. BmaC is a large 340 kDa-protein with a long N-terminal cleavable 72 amino acid signal peptide and several adhesion-related motifs within the passenger domain, an extended pectin lyase virulence factor domain, and several passenger-associated-transport-repeats (PATR) (Figure 1). The portion of BmaC expressed on the phage also exhibits affinity (although much less) for type I collagen, suggesting that this very large protein might contribute to the interaction of *B. suis* with other ECM ligands.

Several lines of evidence have shown that BmaC of *B. suis* 1330 mediates the binding of *B. suis* to epithelial cells through cellular fibronectin. The attachment of *B. suis* to HeLa cells was inhibited by the ϕ-BRA1148 recombinant phage in a dose-dependent manner. A *bmaC* deletion mutant was impaired in the attachment to immobilized fibronectin and to the surface of HeLa and A549 epithelial cells (Table 1). Furthermore, the *bmaC* defective mutant was outcompeted by the wild-type strain in co-infection experiments, and anti-BmaC and anti-fibronectin antibodies significantly inhibited the binding of *B. suis* to HeLa cells [67]. Immunofluorescence microscopy showed that all bacteria with a detectable fluorescent signal displayed BmaC at only one pole, indicating that BmaC is polarly exposed on the cell surface (Figure 2). This is not surprising since several monomeric autotransporters are exposed on the bacterial surface at one pole [76,77]. Confocal microscopy analysis showed the presence of some small GFP-labelled bacterial aggregates on the surface of HeLa cells, mostly in cell boundaries. Single bacteria were found interacting through one of their poles with the cell surface on both the cell body and cell protrusion. Occasionally, polar BmaC was located at the pole interacting with the cell. These observations suggested that the polar localization of BmaC could be relevant in the interaction with host cells in vivo [67].

The monomeric autotransporter proteins encoded by BR0173 and BR2013 (BmaA and BmaB, respectively) of *B. suis* 1330, although much smaller, share significant sequence similarities with BmaC (Figure 1). It was reported that a mutant of *B. suis* 1330, deficient in BmaB (previously called OmaA), is cleared from spleens of BALB/c mice faster than the wild-type strain (between the third and the ninth week post infection), suggesting that BmaB is required during the chronic phase of infection [78]. A recent study [68] indicated that the *bmaB* locus from all *B. abortus* strains analyzed and both the *bmaA* and *bmaC* loci from all *B. melitensis* strains seem to correspond to pseudogenes, while, in *B. suis*, all the Bma proteins could be functional in several strains of this species. In line with these observations, gain or loss of function studies indicated that, at least in *B. suis* strain 1330, BmaA, BmaB, and BmaC proteins contribute, to a greater or lesser degree, to bacterial adhesion among different cell types, such as epithelial (HT-29 and Caco2), synoviocytes, osteoblasts, and trophoblasts (Table 1, Figure 2). These observations show that there are variations in the repertoire of functional adhesins in *Brucella* spp. and open the possibility that these adhesins are involved in host preferences. BmaB was also found at the new pole generated after cell division [68].

5.3.2. Trimeric Autotransporters

As mentioned above, the search for conserved adhesion-associated domains/motifs in *B. suis* 1330 identified a group of trimeric autotransporters, including BR0072 and BR1846, which were named BtaE and BtaF, respectively. The *B. suis* BtaE trimeric autotransporter is a 740 amino acid protein that harbors several regions corresponding to the head and the neck subdomains in addition to a connector region and the β-barrel translocator domain (Figure 1). Different genetic approaches showed that BtaE of *B. suis* is involved in the adhesion to ECM components and host cells [69]. The BtaE-defective strain exhibited a decreased ability to adhere to HeLa and A549 epithelial cells and was outcompeted by the wild-type *B. suis* strain for the adhesion to HeLa cells (Table 1). Expression of BtaE in a "non-adherent" *E. coli* strain increased the binding of this heterologous bacterium to immobilized hyaluronic acid and fibronectin. On the other hand, *btaE* deletion impaired bacterial adhesion to hyaluronic acid but had no effect in the adhesion to fibronectin, suggesting that other fibronectin-binding adhesins (such as BmaC) could compensate somehow for the absence of BtaE. The adhesion of the wild-type strain to HeLa cells decreased in the presence of hyaluronic acid, while this compound had almost no effect in the attachment of the *btaE* mutant to these cells, supporting the hypothesis that BtaE mediates the binding of *Brucella* to hyaluronic acid. Therefore, BtaE could also participate in *Brucella* dissemination to different target tissues such as cartilage, heart, and bone, which may result in brucellosis complications. In vivo experiments using the mouse model indicate that the BtaE adhesin is necessary for a successful infection. In effect, a significantly lower number of bacteria were recovered from spleens of animals inoculated through the intragastric route with the *btaE* mutant compared to those inoculated with the wild-type strain [69].

In a subsequent work, it was shown that the BtaE orthologue of *B. abortus* 2308 is also involved in adhesion to epithelial cells. Compared to *B. suis* BtaE, the ortologue of *B. abortus* is much larger and contains a higher number of repetitive adhesion motifs. Furthermore, the *btaE* gene of *B. suis* and *B. abortus* are under the regulation of different mechanisms (see below) [79]. The *btaE* deletion mutant of *B. abortus* 2308 showed a significant reduction in the adhesion to HeLa cells when compared with the wild-type strain, demonstrating that the BtaE variant of *B. abortus* 2308 contributes to the interaction of *Brucella* with the host cell surface to a similar extent to that observed for the *B. suis* 1330 orthologue [69].

The regulation of *btaE* at a promoter level was analyzed by Sieira et al. [79]. Comparison of *btaE* promoter sequences among different *Brucella* species revealed that a novel HutC binding site in the promoter region of *btaE* from *B. abortus* 2308 was generated de novo recently in the evolution of the genus. HutC, which is a regulator of the histidine metabolism, also acts as a co-activator contributing to modulation of expression of the *Brucella virB* operon [80]. Moreover, additional transcriptional factors (MdrA and IHF) binding sites were identified in the *btaE* promoter of *B. abortus*. The target-DNA sequences were confirmed by EMSA and DNAseI footprinting assays.

The HutC binding site is not present in the *btaE* promoters of other *Brucella* strains since it is interrupted by a cytosine. In effect, an electrophoretic mobility shift assay showed that HutC is not able to interact with the *btaE* promoter of *B. suis* 1330, even though the IHF and MdrA showed a binding pattern similar to that observed for the *btaE* promoter of *B. abortus* 2308. Based on these findings, it was proposed that, as a result of the cis regulatory gain of function, the *btaE* promoter acquired the ability to fine-tune its transcriptional output in response to changes in environmental parameters such as nutrient availability. Thus, differential *btaE* expression might generate phenotypic diversity at the regulatory level of adhesins, which might contribute to reciprocal selection between *Brucella* species and their mammalian hosts.

The BtaF trimeric autotransporter of *B. suis* 1330, encoded by the BR1846 annotated locus, is a small protein that harbors an N-terminal peptide signal, a 170 amino acid passenger domain, and a YadA-like C-terminal translocator region (β-barrel translocator domain) (Figure 1). Unlike BtaE, the BtaF protein does not show the presence of conserved adhesion motifs. Analysis with the TA Domain Annotation (daTAA) server [81] showed that most of the passenger domains correspond to a coiled coil stalk but none associated with the "head" structural region [70]. In addition, a careful analysis of the annotated *btaF* upstream region indicated that the ORF starts earlier, adding 33 additional amino acids at the N-terminal sequence. Using a new version of the program to predict the structural features of trimeric autotransporters and the alternative ORF, it was possible to identify a region at the N-terminus that would correspond to the head, even though the structure would be different from those described so far [82].

BtaF of *B. suis* 1330 has shown to be promiscuous in its ability to bind to different substrates. The heterologous expression of this small trimeric protein markedly increased the adhesion of non-adherent *E. coli* to HeLa cells and various substrates such as fibronectin, fetuin (a sialic acid-rich protein), hyaluronic acid, and collagen I and also to an abiotic surface such as polystyrene [70] (Table 1, Figure 2). In agreement with these observations, the *btaF* deletion mutant of *B. suis* showed a significant reduction in the ability to bind to fetuin, hyaluronic acid, and collagen I, and in the adhesion to an abiotic surface, even though the adhesion to fibronectin was not affected. Again, as it was observed for the *btaE* mutant, overlapping functions with other adhesins (such as Bma proteins and BtaE) may account for the lack of a Δ*btaF* phenotype toward fibronectin. The BtaF-defective strain showed a significant reduction in the attachment to HeLa and A549 epithelial cells.

Notably, BtaF was required for complete virulence in mice infected through the oral route (intragastric administration) [70]. The strain lacking BtaF showed a reduction of about one log in the number of bacteria recovered from spleen at early stages of infection. The absence of both trimeric adhesins (BtaE and BtaF) resulted in a more severe phenotype in vivo compared with the attenuation observed for the single mutants. It is possible that some of the functions might be shared or complementary between BtaE and BtaF, while others could be exclusive to BtaF or BtaE. An indirect ELISA assay on sera from healthy and sick pigs infected with *B. suis* suggested that both adhesins are expressed in vivo in the natural host (swine), supporting the role of these adhesins in the infection process. Recently, it was shown that BtaF is also required for virulence in mice after inoculation via a respiratory (intratracheal administration) route [71]. In this case, the splenic load of the deletion mutant was significantly reduced at 7-days and 30-days post-infection as compared to the wild-type strain.

Smooth *Brucella* strains prevent detection by complement partly due to a distinctive structure of its LPS [83,84]. However, it was proposed that, in addition to the LPS, other surface factors mediate the varied sensitivity of *Brucella* species to the bactericidal action of serum [84]. In addition, it has been shown that, in contrast to human serum, components present in murine normal serum do not opsonize smooth *B. abortus* [85]. The *btaF* mutant showed a significantly reduced survival in the presence of 50% porcine serum compared with the wild-type strain [70]. Furthermore, an *E. coli* strain expressing BtaF showed a more than ten-fold increase in the survival percentage in 8% porcine serum as compared with the control strain. Both strains showed similar levels of survival in heat-inactivated porcine serum, suggesting that BtaF is involved in the resistance to complement-mediated serum killing.

Similar to BmaC and BmaB, BtaE and BtaF adhesins were found to be polarly localized on the bacterial surface [69,70] (Figure 2). Again, the trimeric adhesins were detected in a low proportion of bacteria but, in all cases, the signal showed unipolar localization and, in some cases, sub-polar localization. As it was observed for BmaC and BmaB, they were found at the same pole as AidB-YFP (a new pole marker) [86] and at the opposite of the PdhS-eGFP labeling (old pole marker) [87]. It was proposed that the new pole generated after the asymmetric division would be functionally differentiated for adhesion. An attractive hypothesis is that the initial adhesion of *Brucella* to the host cell would be mediated by adhesins located at the new pole and that adhesin expression only occurs in an infectious bacterial subpopulation. Various cellular mechanisms such as asymmetric division, polar growth, and polar functions generate two functionally differentiated cells [88,89]. Polar localization could be a way of increasing the adhesive power by concentrating the adhesins in a particular region. In fact, host invasion by a bacterial pole can facilitate entry because of the bacterial shape [90]. It is important to note that polar adherence to surfaces is a conserved mechanism shared by several *Alphaproteobacteria* [91,92].

5.3.3. Autotransporters Insertion in the Outer Membrane

Autotransporter translocation into the outer membrane is assisted by the BAM machinery and associated chaperones. More recently, it was shown that the TAM system, made up of TamA and TamB, is also required for the correct insertion of autotransporters from the *Gammaproteobacteria* group (see above) [93].

As mentioned, during the construction of the phylogenetic tree, the BR0049 protein came out as a possible adhesin from the autotransporter family likely due to some structural similarities with this family of proteins. However, our in silico analysis and other reports [94] indicated that BR0049 and its orthologues from other *Alphaproteobacteria* are phylogenetically related to members of the TamB family from *Gammaproteobacteria*. TamB is a large protein mostly periplasmic but inserted in the inner membrane through a non-cleavable signal peptide. BR0049 of *Brucella* spp. shares a relatively low identity (around 22%) with TamB from *Gammaproteobacteria*, but, similarly to this protein, it contains a membrane anchor signal at the N-terminus, which is followed by a region with an abundant β-helix structure, and, at the C-terminus, a short β-barrel structure within the conserved DUF490 domain. As it was proposed for TamB from *Gammaproteobacteria*, it was demonstrated that BR0049 is required for the correct insertion in the OM of the *B. suis* BmaB monomeric autotransporter. In addition, BR0049 was required for complete virulence in mice infected through the intragastric route [53]. The BR0049 mutant showed an increased sensitivity to polymyxin B, lysozyme, and Triton X-100, and, thus, BR0049 was named as MapB (Membrane altering protein). Several results indicated that MapB of *Brucella* plays functions that go beyond that of assisting in autotransporter assembly, suggesting that the TAM machinery would be involved in cell envelope biogenesis.

5.4. *Brucella Adhesins as Vaccine Candidates*

Vaccination is a key health measure in the control and prevention of infectious diseases. In the case of brucellosis, it is necessary to control bacterial dissemination by vaccination of natural hosts as well as vaccination of people professionally exposed to *Brucella* spp. infection. Commercially available *Brucella* vaccines approved for use in animals are based on attenuated strains. These vaccines have serious disadvantages such as producing abortion in pregnant females, being virulent for humans [95–97], and inducing immune responses that interfere with animal serological diagnosis [98]. Thus, there is a need to develop safer and more efficient vaccines [99]. In this sense, acellular vaccines provide great advantages, mainly in terms of safety, not only in its production but also in its administration. Nevertheless, the selection of appropriate vaccine candidates requires the study of host-pathogen interaction.

An essential step in establishing a successful infection is the adhesion of microorganisms to eukaryotic cells, resulting in colonization of the tissue involved. Therefore, the molecules involved in this initial interaction have been widely studied as targets for the development of vaccines against various pathogens such as *E. coli* [100,101], *Haemophilus ducreyi* [102], and *Neisseria meningitidis* [103,104] among others. Despite the extensive knowledge of adhesins' role in the pathogenicity of various bacteria, this group of proteins was not studied well in *Brucella* spp. in terms of immunogenicity and its potential as vaccine candidates.

Al-Mariri et al. studied the immunogenicity and protective efficacy of a DNA vaccine encoding SP41 adhesin from *B. melitensis* in BALB/c mice [105]. Intramuscular (i.m.) administration of pCISP41, a plasmid construct for SP41 expression in mammalian cells, induced SP41-specific serum immunoglobulin G (IgG) antibodies. Moreover, spleen cells from pCISP41-vaccinated mice showed significant T cell proliferation after in vitro stimulation with recombinant SP41 (rSP41) and lysed *B. melitensis*. Splenocytes from pCISP41-immunized animals also responded to rSP41 and bacterial lysate stimulation secreting high levels of gamma interferon (IFN-γ), even though no interleukin-5 (IL-5) was detected. This suggests a predominant T-helper-1 (Th1) immune response.

After an intraperitoneal challenge with *B. melitensis* 16M, mice immunized with pCISP41 exhibited a reduction of 1.25 and 1.14 log in their spleen burden when compared with control mice at 4-weeks and 8-weeks post-challenge, respectively. Nevertheless, vaccination with attenuated *B. melitensis* Rev-1 strain induced better protection levels than pCISP41 vaccination at both time points, achieving a reduction in spleen burden of 1.79 and 3.17 log, respectively. Although SP41 has been shown to be involved in adhesion to epithelial cells, no studies have been made to evaluate the potential of mucosal or systemic vaccination with this adhesin to protect against *Brucella* infection acquired through mucosae.

Brucella infection is frequently acquired through the oral and respiratory routes, and adhesins are anticipated to have a relevant role in these infectious processes. As mentioned above, our results showed that BtaF adhesin from *B. suis* is necessary for complete virulence of *B. suis* after both oral and intratracheal infection [70,71]. In line with these results, we assessed the immunogenic and protective potential of recombinant BtaF when administered intranasally with the mucosal adjuvant Bis-(3′,5′)-cyclic dimeric adenosine monophosphate (c-di-AMP) [71]. To this end, the trimeric form of the BtaF passenger domain fused at the C-terminus to the GCN4tri sequence to facilitate trimerization was successfully expressed and purified [71].

BALB/c mice intranasal immunization with BtaF plus c-di-AMP induced high levels of serum BtaF-specific IgG, IgA, IgG1, and IgG2a, and a mixed IgG1/IgG2a profile. In vitro, these serum antibodies reduced *B. suis* infection of human lung epithelial cells (A549 cell line), reduced bacterial binding to fetuin (a protein rich in sialic acid previously described as a ligand for BtaF), and enhanced *B. suis* phagocytosis by murine macrophages. In addition, immunization led to a significant production of specific IgA antibodies in the airways and in the gastrointestinal and genital mucosae.

BtaF immunization induced a systemic and pulmonary Th1 immune response, shown by the secretion of high levels of IFN-γ by splenocytes and lung cells. Furthermore, depletion of CD4+ or CD8+ populations from spleen cells showed that CD4+ cells were responsible for IFN-γ secretion. BtaF-vaccination also triggered the differentiation of specific CD4+ T cells to central memory cells in cervical lymph nodes, and differentiation of T cells to a Th17 profile in the spleen.

Mice vaccination with BtaF plus c-di-AMP demonstrated a high level of protection against *B. suis* oral infection, reducing the splenic burden of *B. suis* by 3.28 log. Unlike the protection achieved against oral infection, intranasal vaccination with BtaF failed to protect against respiratory infection with *B. suis* since no differences were observed in spleen or lung bacterial load between vaccinated and control mice after an intratracheal challenge.

In summary, despite extensive evidence supporting a role of *Brucella* adhesins in the infectious ability of this pathogen in vivo, only a few studies have assessed the protective efficacy of vaccination with adhesins. Such studies suggest that adhesins hold promise as appropriate antigens for vaccination against oral infection with *Brucella*, even though some protection against systemic infection might be attained.

6. Summary and Future Directions

For a long time, the process of *Brucella* adhesion to host cells has received little attention. Adhesins of *Brucella* spp. identified to date have been studied with varying degrees of depth. Some of them have been analyzed regarding their roles in the binding to components of the ECM and to different cell types as well as their in vivo role, while, in other cases, their functions have been only evaluated in vitro. Still, for various adhesins, it will be necessary to identify their ligands or cell receptors. The possible use of *Brucella* adhesins as vaccine candidates was tested only in two cases, and one of them showed encouraging results. Therefore, the in vivo role of many of the *Brucella* adhesins identified so far, and their possible applications as a basis for acellular vaccines, remains to be evaluated. It is expected that, in the future, new adhesins will be identified that mediate the initial adhesion of *Brucella* to the remarkable variety of cell types that this pathogen can invade. An interesting perspective will be to characterize the roles of the different adhesin variants from different species/strains and to determine whether they play a role in host preferences.

Author Contributions: Conceptualization, P.C.B. and A.Z. Literature review, critical analysis, and synthesis, M.G.B., G.S., M.C.F., F.M.G., P.C.B., and A.Z. Writing—original draft preparation, M.G.B., G.S., M.C.F., F.M.G., P.C.B., and A.Z. Supervision, P.C.B. and A.Z. All authors have read and agreed to the published version of the manuscript.

Funding: This work was supported by the Research Grants from Agencia Nacional de Promoción Científica (PICT 2013-0170, PICT 2016-1199, PICT-2016-2722), and from Universidad de Buenos Aires (UBACYT 20020170100036BA).

Acknowledgments: We are grateful to the staff of the BSL3 facility of INBIRS for expert assistance. We thank current and former members of our laboratories for all the contributions and helpful and inspiring discussions.

Conflicts of Interest: The authors declare no conflict of interest. The funders had no role in the writing of the manuscript, or in the decision to publish the manuscript.

References

1. Pappas, G.; Papadimitriou, P.; Akritidis, N.; Christou, L.; Tsianos, E.V. The new global map of human brucellosis. *Lancet Infect. Dis.* **2006**, *6*, 91–99. [CrossRef]
2. Traxler, R.M.; Lehman, M.W.; Bosserman, E.A.; Guerra, M.A.; Smith, T.L. A Literature Review of Laboratory-Acquired Brucellosis. *J. Clin. Microbiol.* **2013**, *51*, 3055–3062. [CrossRef] [PubMed]
3. Strausbaugh, L.J.; Berkelman, R.L. Human Illness Associated with Use of Veterinary Vaccines. *Clin. Infect. Dis.* **2003**, *37*, 407–414. [CrossRef] [PubMed]
4. Ashford, D.A.; Di Pietra, J.; Lingappa, J.; Woods, C.; Noll, H.; Neville, B.; Weyant, R.; Bragg, S.L.; Spiegel, R.A.; Tappero, J.; et al. Adverse events in humans associated with accidental exposure to the livestock brucellosis vaccine RB51. *Vaccine* **2004**, *22*, 3435–3439. [CrossRef]
5. Blasco, J.M.; Díaz, R. *Brucella melitensis* Rev-1 vaccine as a cause of human brucellosis. *Lancet* **1993**, *342*, 805. [CrossRef]
6. Vincent, P.; Joubert, L.; Prave, M. 2 occupational cases of brucellar infection after inoculation of B 19 vaccine. *Bull. Acad. Vet. Fr.* **1970**, *43*, 89–97.
7. Moreno, E. Retrospective and prospective perspectives on zoonotic brucellosis. *Front. Microbiol.* **2014**, *5*, 213. [CrossRef]
8. Buzgan, T.; Karahocagil, M.K.; Irmak, H.; Baran, A.I.; Karsen, H.; Evirgen, O.; Akdeniz, H. Clinical manifestations and complications in 1028 cases of brucellosis: A retrospective evaluation and review of the literature. *Int. J. Infect. Dis.* **2010**, *14*, e469–e478. [CrossRef]
9. Pourbagher, A.; Pourbagher, M.A.; Savas, L.; Turunc, T.; Demiroglu, Y.Z.; Erol, I.; Yalcintas, D. Epidemiologic, clinical, and imaging findings in brucellosis patients with osteoarticular involvement. *Am. J. Roentgenol.* **2006**, *187*, 873–880. [CrossRef]
10. Godfroid, J.; Cloeckaert, A.; Liautard, J.P.; Kohler, S.; Fretin, D.; Walravens, K.; Garin-Bastuji, B.; Letesson, J.J. From the discovery of the Malta fever's agent to the discovery of a marine mammal reservoir, brucellosis has continuously been a re-emerging zoonosis. *Vet. Res.* **2005**, *36*, 313–326. [CrossRef]
11. Baldi, P.C.; Giambartolomei, G.H. Pathogenesis and pathobiology of zoonotic brucellosis in humans. *OIE Rev. Sci. Tech.* **2013**, *32*, 117–125. [CrossRef] [PubMed]
12. Arenas, G.N.; Staskevich, A.S.; Aballay, A.; Mayorga, L.S. Intracellular trafficking of *Brucella abortus* in J774 macrophages. *Infect. Immun.* **2000**, *68*, 4255–4263. [CrossRef] [PubMed]
13. Price, R.E.; Templeton, J.W.; Smith, R.; Adams, L.G. Ability of mononuclear phagocytes from cattle naturally resistant or susceptible to brucellosis to control in vitro intracellular survival of *Brucella abortus*. *Infect. Immun.* **1990**, *58*, 879–886. [CrossRef]
14. Drazek, E.S.; Houng, H.S.; Crawford, R.M.; Hadfield, T.L.; Hoover, D.L.; Warren, R.L. Deletion of purE attenuates *Brucella melitensis* 16M for growth in human monocyte-derived macrophages. *Infect. Immun.* **1995**, *63*, 3297–3301. [CrossRef]
15. Pizarro-Cerdá, J.; Méresse, S.; Parton, R.G.; Van Der Goot, G.; Sola-Landa, A.; Lopez-Goñi, I.; Moreno, E.; Gorvel, J.P. *Brucella abortus* transits through the autophagic pathway and replicates in the endoplasmic reticulum of nonprofessional phagocytes. *Infect. Immun.* **1998**, *66*, 5711–5724. [CrossRef]
16. Detilleux, P.G.; Deyoe, B.L.; Cheville, N.F. Penetration and intracellular growth of *Brucella abortus* in nonphagocytic cells in vitro. *Infect. Immun.* **1990**, *58*, 2320–2328. [CrossRef] [PubMed]
17. Delpino, M.V.; Fossati, C.A.; Baldi, P.C. Proinflammatory response of human osteoblastic cell lines and Osteoblast-monocyte interaction upon infection with *Brucella* spp. *Infect. Immun.* **2009**, *77*. [CrossRef]
18. Scian, R.; Barrionuevo, P.; Giambartolomei, G.H.; De Simone, E.A.; Vanzulli, S.I.; Fossati, C.A.; Baldi, P.C.; Delpino, M.V. Potential role of fibroblast-like synoviocytes in joint damage induced by *Brucella abortus* infection through production and induction of matrix metalloproteinases. *Infect. Immun.* **2011**, *79*, 3619–3632. [CrossRef]
19. Fernández, A.G.; Ferrero, M.C.; Hielpos, M.S.; Fossati, C.A.; Baldi, P.C. Proinflammatory response of human trophoblastic cells to *Brucella abortus* infection and upon interactions with infected phagocytes. *Biol. Reprod.* **2016**, *94*, 48. [CrossRef] [PubMed]

20. Ferrero, M.C.; Bregante, J.; Delpino, M.V.; Barrionuevo, P.; Fossati, C.A.; Giambartolomei, G.H.; Baldi, P.C. Proinflammatory response of human endothelial cells to *Brucella* infection. *Microbes Infect.* **2011**, *13*, 852–861. [CrossRef]
21. Ferrero, M.C.; Fossati, C.A.; Baldi, P.C. Smooth *Brucella* strains invade and replicate in human lung epithelial cells without inducing cell death. *Microbes Infect.* **2009**, *11*, 476–483. [CrossRef]
22. Billard, E.; Cazevieille, C.; Dornand, J.; Gross, A. High Susceptibility of Human Dendritic Cells to Invasion by the Intracellular Pathogens *Brucella suis*, *B. abortus*, and *B. melitensis*. *Infect. Immun.* **2005**, *73*, 8418–8424. [CrossRef]
23. Delpino, M.V.; Barrionuevo, P.; Scian, R.; Fossati, C.A.; Baldi, P.C. *Brucella*-infected hepatocytes mediate potentially tissue-damaging immune responses. *J. Hepatol.* **2010**, *53*, 145–154. [CrossRef]
24. Ferrero, M.C.; Hielpos, M.S.; Carvalho, N.B.; Barrionuevo, P.; Corsetti, P.P.; Giambartolomei, G.H.; Oliveira, S.C.; Baldi, P.C. Key role of toll-like receptor 2 in the inflammatory response and major histocompatibility complex class ii downregulation in *Brucella abortus*-infected alveolar macrophages. *Infect. Immun.* **2014**, *82*, 626–639. [CrossRef]
25. Fernández, A.G.; Hielpos, M.S.; Ferrero, M.C.; Fossati, C.A.; Baldi, P.C. Proinflammatory response of canine trophoblasts to *Brucella canis* infection. *PLoS ONE* **2017**, *12*, e0186561. [CrossRef]
26. Sidhu-Muñoz, R.S.; Sancho, P.; Vizcaíno, N. Evaluation of human trophoblasts and ovine testis cell lines for the study of the intracellular pathogen *Brucella ovis*. *FEMS Microbiol. Lett.* **2018**, *365*. [CrossRef] [PubMed]
27. Castaneda-Roldan, E.I.; Avelino-Flores, F.; Dall'Agnol, M.; Freer, E.; Cedillo, L.; Dornand, J.; Giron, J.A. Adherence of *Brucella* to human epithelial cells and macrophages is mediated by sialic acid residues. *Cell. Microbiol.* **2004**, *6*, 435–445. [CrossRef]
28. Gomez, G.; Adams, L.G.; Rice-ficht, A.; Ficht, T.A. Host-*Brucella* interactions and the *Brucella* genome as tools for subunit antigen discovery and immunization against brucellosis. *Front. Cell. Infect. Microbiol.* **2013**, *3*, 1–15. [CrossRef]
29. Watarai, M.; Makino, S.I.; Fujii, Y.; Okamoto, K.; Shirahata, T. Modulation of *Brucella*-induced macropinocytosis by lipid rafts mediates intracellular replication. *Cell. Microbiol.* **2002**, *4*, 341–355. [CrossRef] [PubMed]
30. Pei, J.; Turse, J.E.; Ficht, T.A. Evidence of Brucella abortus OPS dictating uptake and restricting NF-k B activation in murine macrophages. *Microbes Infect.* **2008**, *10*, 582–590. [CrossRef]
31. Naroeni, A.; Porte, F. Role of cholesterol and the ganglioside GM1 in entry and short-term survival of Brucella suis in murine macrophages. *Infect. Immun.* **2002**, *70*, 1640–1644. [CrossRef]
32. Kim, S.; Watarai, M.; Suzuki, H.; Makino, S. Lipid raft microdomains mediate class A scavenger receptor-dependent infection of *Brucella abortus*. *Microb. Pathog.* **2004**, *37*, 11–19. [CrossRef]
33. Martín-Martín, A.I.; Vizcaíno, N.; Fernández-Lago, L. Cholesterol, ganglioside GM1 and class A scavenger receptor contribute to infection by *Brucella ovis* and *Brucella canis* in murine macrophages. *Microbes Infect.* **2010**, *12*, 246–251. [CrossRef]
34. Lapaque, N.; Moriyon, I.; Moreno, E.; Gorvel, J.P. *Brucella* lipopolysaccharide acts as a virulence factor. *Curr. Opin. Microbiol.* **2005**, *8*, 60–66. [CrossRef]
35. Manterola, L.; Guzmán-Verri, C.; Chaves-Olarte, E.; Barquero-Calvo, E.; de Miguel, M.-J.; Moriyón, I.; Grilló, M.-J.; López-Goñi, I.; Moreno, E. BvrR/BvrS-controlled outer membrane proteins Omp3a and Omp3b are not essential for *Brucella abortus* virulence. *Infect. Immun.* **2007**, *75*, 4867–4874. [CrossRef]
36. Martín-Martín, A.I.; Caro-Hernández, P.; Orduña, A.; Vizcaíno, N.; Fernández-Lago, L. Importance of the Omp25/Omp31 family in the internalization and intracellular replication of virulent *B. ovis* in murine macrophages and HeLa cells. *Microbes Infect.* **2008**, *10*, 706–710. [CrossRef] [PubMed]
37. Celli, J.; De Chastellier, C.; Franchini, D.M.; Pizarro-Cerda, J.; Moreno, E.; Gorvel, J.P. *Brucella* evades macrophage killing via VirB-dependent sustained interactions with the endoplasmic reticulum. *J. Exp. Med.* **2003**, *198*, 545–556. [CrossRef]
38. Comerci, D.J.; Martínez-Lorenzo, M.J.; Sieira, R.; Gorvel, J.P.; Ugalde, R.A. Essential role of the virB machinery in the maturation of the *Brucella abortus*-containing vacuole. *Cell. Microbiol.* **2001**, *3*, 159–168. [CrossRef]
39. Starr, T.; Child, R.; Wehrly, T.D.; Hansen, B.; Hwang, S.; López-Otin, C.; Virgin, H.W.; Celli, J. Selective subversion of autophagy complexes facilitates completion of the *Brucella* intracellular cycle. *Cell Host Microbe* **2012**, *11*, 33–45. [CrossRef]

40. Porte, F.; Naroeni, A.; Ouahrani-Bettache, S.; Liautard, J.P. Role of the *Brucella suis* lipopolysaccharide O antigen in phagosomal genesis and in inhibition of phagosome-lysosome fusion in murine macrophages. *Infect. Immun.* **2003**, *71*, 1481–1490. [CrossRef] [PubMed]
41. Kline, K.A.; Fälker, S.; Dahlberg, S.; Normark, S.; Henriques-Normark, B. Bacterial Adhesins in Host-Microbe Interactions. *Cell Host Microbe* **2009**, *5*, 580–592. [CrossRef]
42. Pizarro-Cerdá, J.; Cossart, P. Bacterial adhesion and entry into host cells. *Cell* **2006**, *124*, 715–727. [CrossRef]
43. Gerlach, R.G.; Hensel, M. Protein secretion systems and adhesins: The molecular armory of Gram-negative pathogens. *Int. J. Med. Microbiol.* **2007**, *297*, 401–415. [CrossRef]
44. Busch, A.; Waksman, G. Chaperone-usher pathways: Diversity and pilus assembly mechanism. *Philos. Trans. R. Soc. B Biol. Sci.* **2012**, *367*, 1112–1122. [CrossRef]
45. Giltner, C.L.; Nguyen, Y.; Burrows, L.L. Type IV Pilin Proteins: Versatile Molecular Modules. *Microbiol. Mol. Biol. Rev.* **2012**, *76*, 740–772. [CrossRef]
46. Saldaña, Z.; Xicohtencatl-Cortes, J.; Avelino, F.; Phillips, A.D.; Kaper, J.B.; Puente, J.L.; Girón, J.A. Synergistic role of curli and cellulose in cell adherence and biofilm formation of attaching and effacing *Escherichia coli* and identification of Fis as a negative regulator of curli. *Environ. Microbiol.* **2009**, *11*, 992–1006. [CrossRef]
47. Chapman, M.R.; Robinson, L.S.; Pinkner, J.S.; Roth, R.; Heuser, J.; Hammar, M.; Normark, S.; Hultgren, S.J. Role of *Escherichia coli* curli operons in directing amyloid fiber formation. *Science* **2002**, *295*, 851–855. [CrossRef]
48. Delepelaire, P. Type I secretion in gram-negative bacteria. *Biochim. Biophys. Acta Mol. Cell Res.* **2004**, *1694*, 149–161. [CrossRef]
49. Satchell, K.J.F. Structure and function of MARTX toxins and other large repetitive RTX proteins. *Annu. Rev. Microbiol.* **2011**, *65*, 71–90. [CrossRef]
50. Meuskens, I.; Saragliadis, A.; Leo, J.C.; Linke, D. Type V secretion systems: An overview of passenger domain functions. *Front. Microbiol.* **2019**, *10*, 1163. [CrossRef]
51. Babu, M.; Bundalovic-Torma, C.; Calmettes, C.; Phanse, S.; Zhang, Q.; Jiang, Y.; Minic, Z.; Kim, S.; Mehla, J.; Gagarinova, A.; et al. Global landscape of cell envelope protein complexes in *Escherichia coli*. *Nat. Biotechnol.* **2018**, *36*, 103–112. [CrossRef]
52. Kiessling, A.R.; Malik, A.; Goldman, A. Recent advances in the understanding of trimeric autotransporter adhesins. *Med. Microbiol. Immunol.* **2020**, *209*, 233–242. [CrossRef]
53. Bialer, M.G.; Ruiz-Ranwez, V.; Sycz, G.; Estein, S.M.; Russo, D.M.; Altabe, S.; Sieira, R.; Zorreguieta, A. MapB, the *Brucella suis* TamB homologue, is involved in cell envelope biogenesis, cell division and virulence. *Sci. Rep.* **2019**, *9*, 2158. [CrossRef]
54. Albenne, C.; Ieva, R. Job contenders: Roles of the β-barrel assembly machinery and the translocation and assembly module in autotransporter secretion. *Mol. Microbiol.* **2017**, *106*, 505–517. [CrossRef]
55. Van Ulsen, P.; ur Rahman, S.; Jong, W.S.P.; Daleke-Schermerhorn, M.H.; Luirink, J. Type V secretion: From biogenesis to biotechnology. *Biochim. Biophys. Acta Mol. Cell Res.* **2014**, *1843*, 1592–1611. [CrossRef]
56. Emsley, P.; Charles, I.G.; Fairweather, N.F.; Isaacs, N.W. Structure of *Bordetella pertussis* virulence factor P.69 pertactin. *Nature* **1996**, *381*, 90–92. [CrossRef]
57. Oberhettinger, P.; Leo, J.C.; Linke, D.; Autenrieth, I.B.; Schütz, M.S. The inverse autotransporter intimin exports its passenger domain via a hairpin intermediate. *J. Biol. Chem.* **2015**, *290*, 1837–1849. [CrossRef]
58. Leo, J.C.; Oberhettinger, P.; Schütz, M.; Linke, D. The inverse autotransporter family: Intimin, invasin and related proteins. *Int. J. Med. Microbiol.* **2015**, *305*, 276–282. [CrossRef]
59. Leibiger, K.; Schweers, J.M.; Schütz, M. Biogenesis and function of the autotransporter adhesins YadA, intimin and invasin. *Int. J. Med. Microbiol.* **2019**, *309*, 331–337. [CrossRef]
60. del Rocha-Gracia, R.C.; Castañeda-Roldán, E.I.; Giono-Cerezo, S.; Girón, J.A. *Brucella sp.* bind to sialic acid residues on human and animal red blood cells. *FEMS Microbiol. Lett.* **2002**, *213*, 219–224. [CrossRef]
61. Castañeda-Roldán, E.I.; Ouahrani-Bettache, S.; Saldaña, Z.; Avelino, F.; Rendón, M.A.; Dornand, J.; Girón, J.A. Characterization of SP41, a surface protein of *Brucella* associated with adherence and invasion of host epithelial cells. *Cell. Microbiol.* **2006**, *8*, 1877–1887. [CrossRef] [PubMed]
62. Sidhu-Muñoz, R.S.; Sancho, P.; Vizcaíno, N. Brucella ovis PA mutants for outer membrane proteins Omp10, Omp19, SP41, and BepC are not altered in their virulence and outer membrane properties. *Vet. Microbiol.* **2016**, *186*, 59–66. [CrossRef]

63. Czibener, C.; Ugalde, J.E. Identification of a unique gene cluster of *Brucella* spp. that mediates adhesion to host cells. *Microbes Infect.* **2012**, *14*, 79–85. [CrossRef]
64. Czibener, C.; Merwaiss, F.; Guaimas, F.; Del Giudice, M.G.; Serantes, D.A.R.; Spera, J.M.; Ugalde, J.E. BigA is a novel adhesin of *Brucella* that mediates adhesion to epithelial cells. *Cell. Microbiol.* **2016**, *18*, 500–513. [CrossRef]
65. Lopez, P.; Guaimas, F.; Czibener, C.; Ugalde, J.E. A genomic island in *Brucella* involved in the adhesion to host cells: Identification of a new adhesin and a translocation factor. *Cell. Microbiol.* **2020**, e13245. [CrossRef]
66. Eltahir, Y.; Al-Araimi, A.; Nair, R.; Autio, K.J.; Tu, H.; Leo, J.C.; Al-Marzooqi, W.; Johnson, E. Binding of *Brucella* protein, Bp26, to select extracellular matrix molecules. *BMC Mol. Cell Biol.* **2019**, *20*, 55. [CrossRef]
67. Posadas, D.M.; Ruiz-Ranwez, V.; Bonomi, H.R.; Martín, F.A.; Zorreguieta, A. BmaC, a novel autotransporter of *Brucella suis*, is involved in bacterial adhesion to host cells. *Cell. Microbiol.* **2012**, *14*, 965–982. [CrossRef]
68. Bialer, M.G.; Ferrero, M.C.; Delpino, M.V.; Ruiz-Ranwez, V.; Posadas, D.M.; Baldi, P.C.; Zorreguieta, A. Adhesive functions or pseudogenization of monomeric autotransporters in *Brucella* species. Unpublished.
69. Ruiz-Ranwez, V.; Posadas, D.M.; Van der Henst, C.; Estein, S.M.; Arocena, G.M.; Abdian, P.L.; Martín, F.A.; Sieira, R.; De Bolle, X.; Zorreguieta, A. BtaE, an Adhesin That Belongs to the Trimeric Autotransporter Family, Is Required for Full Virulence and Defines a Specific Adhesive Pole of *Brucella suis*. *Infect. Immun.* **2013**, *81*, 996–1007. [CrossRef]
70. Ruiz-Ranwez, V.; Posadas, D.M.; Estein, S.M.; Abdian, P.L.; Martin, F.A.; Zorreguieta, A. The BtaF trimeric autotransporter of *Brucella suis* is involved in attachment to various surfaces, resistance to serum and virulence. *PLoS ONE* **2013**, *8*, e79770. [CrossRef] [PubMed]
71. Muñoz González, F.; Sycz, G.; Alonso Paiva, I.M.; Linke, D.; Zorreguieta, A.; Baldi, P.C.; Ferrero, M.C. The BtaF Adhesin Is Necessary for Full Virulence During Respiratory Infection by *Brucella suis* and Is a Novel Immunogen for Nasal Vaccination Against Brucella Infection. *Front. Immunol.* **2019**, *10*, 1775. [CrossRef]
72. Vitry, M.A.; Mambres, D.H.; Deghelt, M.; Hack, K.; Machelart, A.; Lhomme, F.; Vanderwinden, J.M.; Vermeersch, M.; De Trez, C.; Pérez-Morga, D.; et al. *Brucella melitensis* invades murine erythrocytes during infection. *Infect. Immun.* **2014**, *82*, 3927–3938. [CrossRef] [PubMed]
73. Seco-Mediavilla, P.; Verger, J.M.; Grayon, M.; Cloeckaert, A.; Marín, C.M.; Zygmunt, M.S.; Fernández-Lago, L.; Vizcaíno, N. Epitope mapping of the *Brucella melitensis* BP26 immunogenic protein: Usefulness for diagnosis of sheep brucellosis. *Clin. Diagn. Lab. Immunol.* **2003**, *10*, 647–651. [CrossRef]
74. Bodelón, G.; Palomino, C.; Fernández, L.Á. Immunoglobulin domains in *Escherichia coli* and other enterobacteria: From pathogenesis to applications in antibody technologies. *FEMS Microbiol. Rev.* **2013**, *37*, 204–250. [CrossRef]
75. Posadas, D.M. Transport and Adhesion in *Brucella suis*: Role of a TolC Family Protein in the Efflux of Toxic Compounds and of Three Potential Adhesins Host in Colonization. Ph.D. Thesis, University of Buenos Aires, Buenos Aires, Argentina, 2010.
76. Jain, S.; Van Ulsen, P.; Benz, I.; Schmidt, M.A.; Fernandez, R.; Tommassen, J.; Goldberg, M.B. Polar localization of the autotransporter family of large bacterial virulence proteins. *J. Bacteriol.* **2006**, *188*, 4841–4850. [CrossRef]
77. Goldberg, M.B.; Barzu, O.; Parsot, C.; Sansonetti, P.J. Unipolar localization and ATPase activity of IcsA, a *Shigella flexneri* protein involved in intracellular movement. *J. Bacteriol.* **1993**, *175*, 2189–2196. [CrossRef]
78. Bandara, A.B.; Sriranganathan, N.; Schurig, G.G.; Boyle, S.M. Putative outer membrane autotransporter protein influences survival of *Brucella suis* in BALB/c mice. *Vet. Microbiol.* **2005**, *109*, 95–104. [CrossRef]
79. Sieira, R.; Bialer, M.G.; Roset, M.S.; Ruiz-Ranwez, V.; Langer, T.; Arocena, G.M.; Mancini, E.; Zorreguieta, A. Combinatorial control of adhesion of *Brucella abortus* 2308 to host cells by transcriptional rewiring of the trimeric autotransporter btaE gene. *Mol. Microbiol.* **2017**, *103*, 553–565. [CrossRef]
80. Sieira, R.; Arocena, G.M.; Bukata, L.; Comerci, D.J.; Ugalde, R.A. Metabolic control of virulence genes in *Brucella abortus*: HutC coordinates virB expression and the histidine utilization pathway by direct binding to both promoters. *J. Bacteriol.* **2010**, *192*, 217–224. [CrossRef]
81. Szczesny, P.; Lupas, A. Domain annotation of trimeric autotransporter adhesins—daTAA. *Bioinformatics* **2008**, *24*, 1251–1256. [CrossRef]
82. Linke, D. (University of Oslo, Oslo, Norway). Personal communication, 2018.
83. Eisenschenk, F.C.; Houle, J.J.; Hoffmann, E.M. Mechanism of serum resistance among *Brucella abortus* isolates. *Vet. Microbiol.* **1999**, *68*, 235–244. [CrossRef]

84. Fernandez-Prada, C.M.; Nikolich, M.; Vemulapalli, R.; Sriranganathan, N.; Boyle, S.M.; Schurig, G.G.; Hadfield, T.L.; Hoover, D.L. Deletion of wboA enhances activation of the lectin pathway of complement in Brucella abortus and *Brucella melitensis*. *Infect. Immun.* **2001**, *69*, 4407–4416. [CrossRef]
85. Mora-Cartín, R.; Chacón-Díaz, C.; Gutiérrez-Jiménez, C.; Gurdián-Murillo, S.; Lomonte, B.; Chaves-Olarte, E.; Barquero-Calvo, E.; Moreno, E. N -Formyl-Perosamine Surface Homopolysaccharides Hinder the Recognition of *Brucella abortus* by Mouse Neutrophils. *Infect. Immun.* **2016**, *84*, 1712–1721. [CrossRef]
86. Dotreppe, D.; Mullier, C.; Letesson, J.J.; De Bolle, X. The alkylation response protein AidB is localized at the new poles and constriction sites in *Brucella abortus*. *BMC Microbiol.* **2011**, *11*, 257. [CrossRef]
87. Hallez, R.; Mignolet, J.; Van Mullem, V.; Wery, M.; Vandenhaute, J.; Letesson, J.J.; Jacobs-Wagner, C.; De Bolle, X. The asymmetric distribution of the essential histidine kinase PdhS indicates a differentiation event in *Brucella abortus*. *EMBO J.* **2007**, *26*, 1444–1455. [CrossRef] [PubMed]
88. Brown, P.J.B.; De Pedro, M.A.; Kysela, D.T.; Van Der Henst, C.; Kim, J.; De Bolle, X.; Fuqua, C.; Brun, Y.V. Polar growth in the Alphaproteobacterial order Rhizobiales. *Proc. Natl. Acad. Sci. USA* **2012**, *109*, 1697–1701. [CrossRef]
89. Hallez, R.; Bellefontaine, A.F.; Letesson, J.J.; De Bolle, X. Morphological and functional asymmetry in α-proteobacteria. *Trends Microbiol.* **2004**, *12*, 361–365. [CrossRef]
90. Van Der Henst, C.; De Barsy, M.; Zorreguieta, A.; Letesson, J.J.; De Bolle, X. The *Brucella* pathogens are polarized bacteria. *Microbes Infect.* **2013**, *15*, 998–1004. [CrossRef]
91. Tomlinson, A.D.; Fuqua, C. Mechanisms and regulation of polar surface attachment in *Agrobacterium tumefaciens*. *Curr. Opin. Microbiol.* **2009**, *12*, 708–714. [CrossRef]
92. Merker, R.I.; Smit, J. Characterization of the Adhesive Holdfast of Marine and Freshwater Caulobacters. *Appl. Environ. Microbiol.* **1988**, *54*, 2078–2085. [CrossRef]
93. Selkrig, J.; Mosbahi, K.; Webb, C.T.; Belousoff, M.J.; Perry, A.J.; Wells, T.J.; Morris, F.; Leyton, D.L.; Totsika, M.; Phan, M.D.; et al. Discovery of an archetypal protein transport system in bacterial outer membranes. *Nat. Struct. Mol. Biol.* **2012**, *19*, 506–510. [CrossRef] [PubMed]
94. Heinz, E.; Selkrig, J.; Belousoff, M.J.; Lithgow, T. Evolution of the translocation and assembly module (TAM). *Genome Biol. Evol.* **2015**, *7*, 1628–1643. [CrossRef]
95. Davis, D.S.; Templeton, J.W.; Ficht, T.A.; Huber, J.D.; Angus, R.D.; Adams, L.G. *Brucella abortus* in Bison. II. Evaluation of strain 19 vaccination of pregnant cows. *J. Wildl. Dis.* **1991**, *27*, 258–264. [CrossRef] [PubMed]
96. Spink, W.W.; Hall, J.W.; Finstad, J.; Mallet, E. Immunization with viable *Brucella* organisms: Results of a safety test in humans. *Bull. World Health Organ.* **1962**, *26*, 409–419.
97. Pappagianis, D.; Elberg, S.S.; Crouch, D. Immunization against brucella infections: Effects of graded doses of viable attenuated *Brucella melitensis* in humans. *Am. J. Epidemiol.* **1966**, *84*, 21–31. [CrossRef]
98. Schurig, G.G.; Sriranganathan, N.; Corbel, M.J. Brucellosis vaccines: Past, present and future. *Vet. Microbiol.* **2002**, *90*, 479–496. [CrossRef]
99. Neutra, M.R.; Kozlowski, P.A. Mucosal vaccines: The promise and the challenge. *Nat. Rev. Immunol.* **2006**, *6*, 148–158. [CrossRef]
100. Pecha, B.; Low, D.; O'Hanley, P. Gal-Gal pili vaccines prevent pyelonephritis by piliated *Escherichia coli* in a murine model. Single-component Gal-Gal pili vaccines prevent pyelonephritis by homologous and heterologous piliated *E. coli* strains. *J. Clin. Investig.* **1989**, *83*, 2102–2108. [CrossRef]
101. Langermann, S.; Palaszynski, S.; Barnhart, M.; Auguste, G.; Pinkner, J.S.; Burlein, J.; Barren, P.; Koenig, S.; Leath, S.; Jones, C.H.; et al. Prevention of mucosal *Escherichia coli* infection by FimH-adhesin-based systemic vaccination. *Science* **1997**, *276*, 607–611. [CrossRef]
102. Samo, M.; Choudhary, N.R.; Riebe, K.J.; Shterev, I.; Staats, H.F.; Sempowski, G.D.; Leduc, I. Immunization with the *Haemophilus ducreyi* trimeric autotransporter adhesin DsrA with alum, CpG or imiquimod generates a persistent humoral immune response that recognizes the bacterial surface. *Vaccine* **2016**, *34*, 1193–1200. [CrossRef] [PubMed]
103. Comanducci, M.; Bambini, S.; Brunelli, B.; Adu-Bobie, J.; Aricò, B.; Capecchi, B.; Giuliani, M.M.; Masignani, V.; Santini, L.; Savino, S.; et al. NadA, a novel vaccine candidate of *Neisseria meningitidis*. *J. Exp. Med.* **2002**, *195*, 1445–1454. [CrossRef]

104. Hung, M.C.; Heckels, J.E.; Christodoulides, M. The adhesin complex protein (ACP) of *Neisseria meningitidis* is a new adhesin with vaccine potential. *MBio* **2013**, *4*, e00041-13. [CrossRef] [PubMed]
105. Al-Mariri, A.; Abbady, A.Q. Evaluation of the immunogenicity and the protective efficacy in mice of a DNA vaccine encoding SP41 from *Brucella melitensis*. *J. Infect. Dev. Ctries.* **2013**, *7*, 329–337. [CrossRef]

pathogens MDPI

Article

ASC-Mediated Inflammation and Pyroptosis Attenuates *Brucella abortus* Pathogenesis Following the Recognition of gDNA

Juselyn D. Tupik [1], Sheryl L. Coutermarsh-Ott [1], Angela H. Benton [1], Kellie A. King [1], Hanna D. Kiryluk [1], Clayton C. Caswell [1] and Irving C. Allen [1,2,*]

[1] Department of Biomedical Sciences and Pathobiology, Virginia-Maryland College of Veterinary Medicine, Virginia Tech, Blacksburg, VA 24061, USA; jdtupik@vt.edu (J.D.T.); slc2003@vt.edu (S.L.C.-O.); ahbenton@vt.edu (A.H.B.); kellieking@vt.edu (K.A.K.); hanna98@vt.edu (H.D.K.); caswellc@vt.edu (C.C.C.)

[2] Department of Basic Science Education, Virginia Tech Carilion School of Medicine, Roanoke, VA 24016, USA

* Correspondence: icallen@vt.edu; Tel.: +1-540-231-7551; Fax: 1-540-231-6033

Received: 31 October 2020; Accepted: 25 November 2020; Published: 30 November 2020

Abstract: *Brucella abortus* is a zoonotic pathogen that causes brucellosis. Because of *Brucella's* unique LPS layer and intracellular localization predominately within macrophages, it can often evade immune detection. However, pattern recognition receptors are capable of sensing *Brucella* pathogen-associated molecular patterns (PAMPS). For example, NOD-like receptors (NLRs) can form a multi-protein inflammasome complex to attenuate *Brucella* pathogenesis. The inflammasome activates IL-1β and IL-18 to drive immune cell recruitment. Alternatively, inflammasome activation also initiates inflammatory cell death, termed pyroptosis, which augments bacteria clearance. In this report, we assess canonical and non-canonical inflammasome activation following *B. abortus* infection. We conducted in vivo studies using $Asc^{-/-}$ mice and observed decreased mouse survival, immune cell recruitment, and increased bacteria load. We also conducted studies with $Caspase\text{-}11^{-/-}$ mice and did not observe any significant impact on *B. abortus* pathogenesis. Through mechanistic studies using $Asc^{-/-}$ macrophages, our data suggests that the protective role of ASC may result from the induction of pyroptosis through a gasdermin D-dependent mechanism in macrophages. Additionally, we show that the recognition of *Brucella* is facilitated by sensing the PAMP gDNA rather than the less immunogenic LPS. Together, these results refine our understanding of the role that inflammasome activation and pyroptosis plays during brucellosis.

Keywords: brucellosis; canonical inflammasome; non-canonical inflammasome; NLR; pyroptosis; ASC; caspase-11; caspase-1; IL-1β; gDNA

1. Introduction

Brucellosis is a zoonotic bacterial disease that exhibits pathogenesis consistent with inflammation. Transmitted through *Brucella* spp. primarily from agricultural animals to humans in unpasteurized dairy products, brucellosis symptoms in humans often include inflammatory or influenza-like characteristics such as arthritis, undulant fevers, and neurological manifestations [1,2]. Because there is currently no human vaccine for brucellosis and effective antibiotic regiments for the disease require long treatment durations, these symptoms often persist throughout the infected individual's lifetime due to the well-adapted ability of *Brucella* to evade immune recognition [3]. Unlike the classical lipopolysaccharide (LPS) layer of Gram-negative bacteria such as *Escherichia coli* that contain a glucosamine backbone with short acyl groups, *Brucella* spp. contain a modified lipid A layer that consists of a diaminoglucose backbone with long branching acyl groups [2]. This deviation from a consistent molecular structure has the potential to subvert immune recognition by the innate immune system through complement interference and decreased cytokine production, leading to enhanced *Brucella* replication and pathogenesis. Despite its mechanisms of immune avoidance, there are some aspects of *Brucella* spp. that are recognized by the innate immune system, making the understanding of these mechanisms essential for targeting future treatments for brucellosis.

As hypothesized by Janeway (1989), the innate immune system has evolved over time to recognize consistent molecular structures in pathogens known as Pathogen or Damage-Associated Molecular Patterns (PAMPs or DAMPs). These PAMPs and DAMPs are recognized by protein structures known as pattern recognition receptors (PRRs) [4]. From previous studies, *Brucella* genomic DNA (gDNA) is known to be recognized by the PRR absent in melanoma 2 (AIM2) and subsequently promotes inflammation, making it an excellent PAMP for immune recognition [5–7]. PRRs include membrane-bound receptors, which consist of Toll-like receptors (TLRs) and C-type Lectin receptors (CLRs), as well as cytosolic receptors made up of a Nucleotide-Binding Domain and Leucine-Rich Repeat Containing receptors (NLRs), Aim-2-Like receptors (ALRs), Rig-I-Like Helicase receptors (RLRs), and the X-LR class of uncategorized receptors [8,9]. After the recognition of a PAMP or DAMP, PRRs generally serve as scaffolding proteins to promote the initiation or inhibition of immune signaling pathways [9]. Of the PRRs that have been described in brucellosis, the best characterized have been the TLRs. From previous literature, many TLRs have been implicated with *Brucella* detection, which plays a role in bacterial signaling, host resistance, and dendritic cell activation [10–16]. TLRs also play important roles in the transcriptional generation of inactive inflammatory cytokines in response to *Brucella* infections that can be activated by NLR or ALR immune signaling complexes [17,18]. This indicates that multiple PRRs work in tandem to attenuate brucellosis pathogenesis.

Inflammasome-forming NLRs and AIM2 have also been reported to play a role in *Brucella* sensing [5–7]. After recognition, the NLR or ALR is able to bind the apoptosis-associated speck-like protein containing a caspase activation recruitment domain (CARD) (ASC) and procaspase-1 to form the canonical inflammasome [8]. The inflammasome then cleaves caspase-1, which subsequently cleaves the cytokines pro-IL-1β and pro-IL-18, produced through TLR signaling, to their active forms to promote inflammation [8,9,19–23]. Inflammasome signaling can also lead to a form of inflammatory cell death known as pyroptosis. Pyroptosis occurs when activated capase-1 cleaves the protein gasdermin D, releasing the gasdermin N subunit [24]. This subunit binds with phosphoinositides on the cell membrane and oligomerizes, creating membrane pores that lead to an osmotic imbalance in the cell that eventually leads to cell lysis [24]. Recently, the formation of a non-canonical inflammasome has been described that utilizes caspase-11 to mediate the cleavage of gasdermin D to initiate pyroptosis.

Previous studies evaluating inflammasome activation in response to *Brucella* have predominately focused on characterizing the activation of inflammatory cytokine signaling associated with canonical inflammasome activation. The best described inflammasomes involved in *Brucella* infections are NLR Family Pyrin Domain Containing 3 (NLRP3) and AIM2. In mouse models, the NLRP3 inflammasome promotes survival and decreased bacterial load through enhanced cytokine secretion, in addition to sensing mitochondrial reactive oxygen species (ROS) generated from *Brucella* [6,7]. Looking at the *Brucella* PAMPs, AIM2, as a known sensor of bacterial DNA, becomes activated from *Brucella* gDNA recognition and initiates inflammatory cytokine signaling and pyroptosis [5,6,25]. These inflammasomes are ASC-dependent, as shown in the formation of punctate ASC structures during infection [6], indicating that ASC-dependent inflammasomes are important in *Brucella* recognition and targeting through inflammatory cytokine signaling. Despite these advancements in studying inflammasome-mediated inflammatory cytokine signaling, pyroptosis and the role of gasdermin D have not been extensively evaluated in response to *Brucella*.

In this study, we used *Asc*$^{-/-}$ and *Caspase-11*$^{-/-}$ mice to further elucidate the role of the canonical and non-canonical inflammasomes following *B. abortus* infection. We sought to assess survival, histopathology, bacterial load, and cell death to provide a more holistic view of cytokine responses and pyroptosis, both in vivo and *in vitro*. Additionally, we reassessed *Brucella* PAMPs using *B. abortus* gDNA and LPS to better define the mechanisms associated with pathogen recognition. Ultimately, we found that ASC functions to attenuate *B. abortus* pathogenesis through the modulation of inflammation and pyroptosis, requiring gasdermin D through a mechanism independent of caspase-11. Additionally, we determined that *Brucella* gDNA, rather than LPS, provoked an elevated inflammasome response that augmented pyroptosis. This report contributes to the current literature and provides some additional novel insights into potential mechanisms of inflammasome activation during brucellosis.

2. Results

2.1. ASC Attenuates B. abortus Pathogenesis and Is Critical for Host Survival

To explore canonical and non-canonical inflammasome activation to *Brucella abortus*, we used mice that lack either the ASC adaptor protein ($Asc^{-/-}$) or the non-canonical inflammasome-associated caspase, caspase-11 ($Caspase\text{-}11^{-/-}$). After intraperitoneally injecting mice with 1×10^5 colony forming units (CFUs) of *B. abortus*, we monitored mortality in all mouse groups for a 24-day period (Figure 1A). At Day 7, there was a 26.3% decrease in the survival of the $Asc^{-/-}$ mice. In these animals, the weight loss in 5 of the 19 mice exceeded 20%, and several of the $Asc^{-/-}$ mice developed clinical parameters associated with disease progression, such as decreased body condition, that required euthanasia (Figure 1B). However, there was no decrease in survival for the wildtype (WT) and $Caspase\text{-}11^{-/-}$ mice. These mortality data suggest that the canonical inflammasome plays a more critical role in host survival compared to the non-canonical inflammasome and caspase-11.

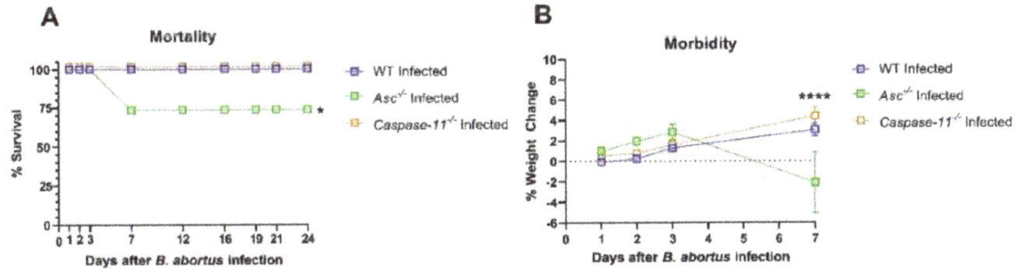

Figure 1. $Asc^{-/-}$ and $Caspase\text{-}11^{-/-}$ mortality and morbidity. $Asc^{-/-}$ ($n = 19$), $Caspase\text{-}11^{-/-}$ ($n = 19$), and C57BL/6 WT ($n = 26$) mice were injected intraperitoneally with 1×10^5 *B. abortus* CFUs and assessed daily for excessive weight loss (>20%) warranting euthanasia according to the Institutional Animal Care and Use Committee (IACUC). (**A**) Mortality graph based on (**B**) a morbidity assessment of $Asc^{-/-}$, $Capsase\text{-}11^{-/-}$, and WT mice. All the mice were weighed for a 24-day period, with morbidity warranting euthanasia in $Asc^{-/-}$ mice only at Day 7. * $p < 0.05$, **** $p < 0.0001$.

2.2. ASC Contributes to Inflammatory Pathogenesis during B. abortus Infection

To further elucidate the role of ASC and caspase-11 activation *in vivo*, we conducted a histopathological analysis on the liver and spleen from wildtype, $Asc^{-/-}$, and $Caspase\text{-}11^{-/-}$ mice three days post-infection. Histopathology indicated that all the infected mouse groups exhibited elevated extramedullary hematopoiesis (EMH) and inflammation in the liver and spleen (Figure 2A,B).

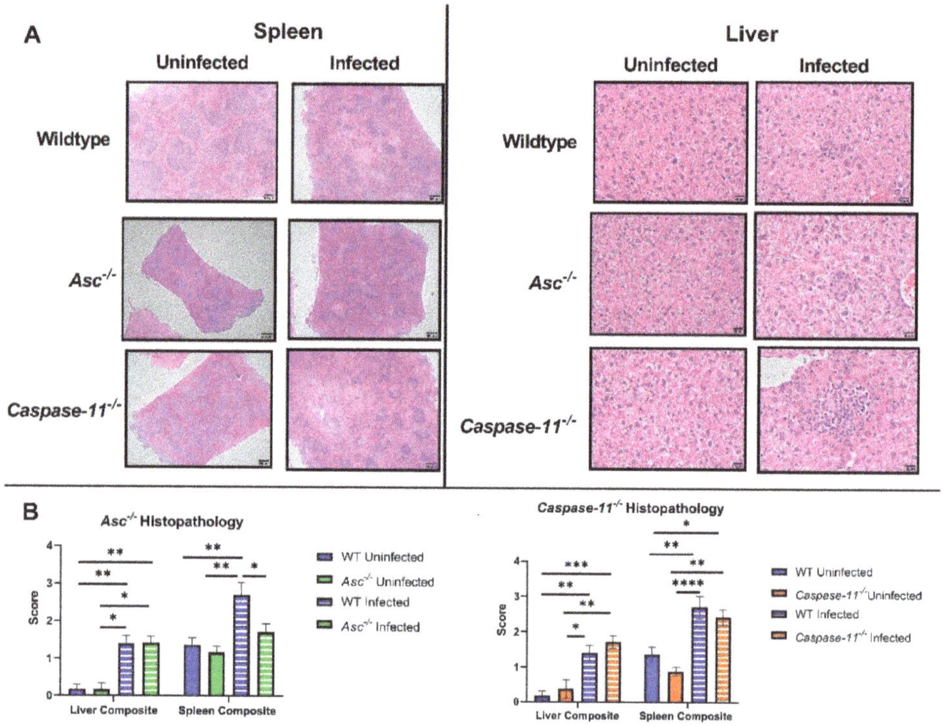

Figure 2. Inflammation in the liver and spleen. A total of 20 *Asc*$^{-/-}$ (n = 6 uninfected [U], 14 infected [I]), 27 *Caspase-11*$^{-/-}$ (n = 8 U, 19 I), and 34 C57BL/6 WT (n = 11 U, 23 I) livers and spleens were evaluated by histopathology 3 d.p.i. (**A**) H&E stained histological slides of the liver and spleen from WT, *Asc*$^{-/-}$, and *Caspase-11*$^{-/-}$ mice. All spleen images were taken at 4× power and all liver images were taken at 40× power. Inflammation and extramedullary hematopoiesis (EMH) were the dominant features observed in the histopathology evaluation. (**B**) Bar graphs of WT, *Asc*$^{-/-}$, and *Caspase-11*$^{-/-}$ histopathology composite scores were generated based on inflammation and EMH in the liver and spleen. * $p < 0.05$, ** $p < 0.01$, *** $p < 0.001$, **** $p < 0.0001$.

Wildtype mice exhibited higher EMH and inflammation scores compared to the *Asc*$^{-/-}$ animals in the spleen. This trend was not observed in the *Caspase-11*$^{-/-}$ mice in either the liver or the spleen. Together, these data suggest that ASC augments splenic inflammation and plays a key role in *B. abortus*-mediated pathology in the spleen, which is a target organ in this mouse model. The failure to mount a vigorous immune response to *B. abortus* in the *Asc*$^{-/-}$ mice is likely associated with the increased morbidity and mortality.

2.3. Bacterial Load Is Significantly Increased in the Absence of ASC and Decreased in the Absence of Caspase-11

Bacterial loads were determined in the spleen and liver of *B. abortus*-infected animals 3 d.p.i (Figure 3). Between wildtype, *Asc*$^{-/-}$, and *Caspase-11*$^{-/-}$ mice, there was no significant difference in the weight of the spleens used for analysis.

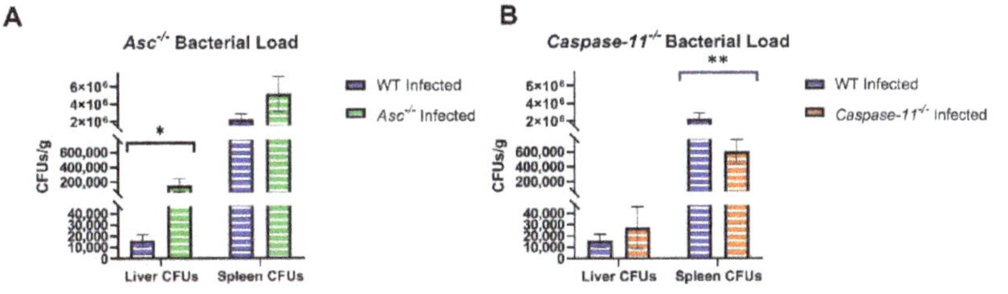

Figure 3. *Brucella* CFUs in the liver and spleen of $Asc^{-/-}$ and $Caspase\text{-}11^{-/-}$ mice. Three d.p.i. liver and spleens from infected C57BL/6 WT ($n = 23$), **(A)** $Asc^{-/-}$ ($n = 14$), and **(B)** $Caspase\text{-}11^{-/-}$ ($n = 19$) mice were homogenized and counted for CFUs/gram. * $p < 0.05$, ** $p < 0.01$.

In $Asc^{-/-}$ mice, we observed significantly elevated *B. abortus* CFUs in the liver and a trending increase in the spleen. In the $Caspase\text{-}11^{-/-}$ mice, we observed a similar trending increase in CFUs in the liver. However, in the spleen $Caspase\text{-}11^{-/-}$ mice had significantly decreased bacterial CFUs compared to the wildtype. The increased bacteria load in $Asc^{-/-}$ mice is consistent with the increased morbidity and reduced inflammation in Figures 1 and 2, and further illustrates the critical role of ASC in the host response to *B. abortus*. Likewise, these $Caspase\text{-}11^{-/-}$ data suggest a significant, but variable, role in controlling the *B. abortus* bacteria burden that does not appear to impact the overall host morbidity or inflammation.

2.4. *B. abortus* Initiates a Weak Inflammasome-Mediated Inflammatory Cytokine Response

Inflammasome activation results in the cleavage and processing of IL-1β and IL-18. To evaluate the generation of pro-IL-1β, so-called "Signal 1", at the transcript level, we conducted quantitative real-time PCR in liver and spleen homogenates (Figure 4). We observed significantly increased fold changes in *Il1β* in the livers of $Asc^{-/-}$ and wildtype mice infected with *B. abortus* versus the uninfected mice (Figure 4A). Within infected mouse groups in the liver, we also found that infected $Asc^{-/-}$ mice had a significantly decreased fold change of *Il1β* compared to wildtype mice (Figure 4A). However, when assessing the total IL-1β (uncleaved and cleaved) protein, we did not see a significant difference between the infected wildtype and $Asc^{-/-}$ mice.

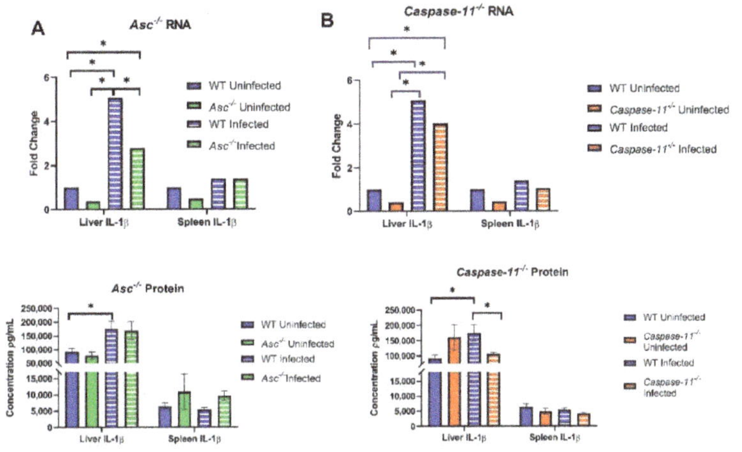

Figure 4. Inflammatory signaling in the liver and spleen of $Asc^{-/-}$ and $Caspase\text{-}11^{-/-}$ mice. Three d.p.i. livers and spleens of 34 C57BL/6 WT ($n = 11$ U, 23 I), (**A**) 20 $Asc^{-/-}$ ($n = 6$ U, 14 I), and (**B**) 27 $Caspase\text{-}11^{-/-}$ ($n = 8$ U, 19 I) mice were homogenized and analyzed for their *Il1β* RNA fold change using RT-PCR and IL-1β protein concentration through ELISA. * $p < 0.05$.

In experiments with *Caspase-11*$^{-/-}$ mice, we also found an elevated RNA fold change between the infected and uninfected groups (Figure 4B), but there was no significant difference between the infected wildtype and *Caspase-11*$^{-/-}$ mice for *Il1β* transcription (Figure 4B). However, there was a minimal, but statistically significant, decrease in IL-1β protein in the livers from *Caspase-11*$^{-/-}$ mice compared to the wildtype animals.

2.5. *B. abortus* Infection Attenuated IL-1β and Induced a Strong ASC-Dependent Pyroptosis Response in Macrophages

Due to the significant phenotype observed in the *Asc*$^{-/-}$ mice, we next sought to better define the underlying mechanism using ex vivo bone marrow-derived macrophages (BMDMs). Intracellular bacterial replication and survival was evaluated over a 48 h period in *B. abortus*-infected *Asc*$^{-/-}$ and wildtype BMDMs. Under these conditions, we observed a significant decrease in *B. abortus* growth in the *Asc*$^{-/-}$ macrophages compared with the wildtype BMDMs after 24 h (Figure 5A). By 48 h, the *B. abortus* replication and survival was no longer detectable, while increasing in the wildtype BMDMs. These results were unexpected based on our in vivo findings and suggest a possible disconnect between pathogen clearance, inflammasome function, and pyroptosis in BMDMs.

We next measured the RNA fold change and protein concentration of IL-1β (Figure 5B,C). Within 2 h of *B. abortus* exposure, we observed elevated *Il1β* transcription in the infected wildtype and *Asc*$^{-/-}$ macrophages. Transcription was statistically significant, but only slightly higher, in the *Asc*$^{-/-}$ cells compared to the wildtype BMDMs (Figure 5B). At 24 h and 48 h, we observed a significant decrease in *Il1β* transcription in both groups of infected mice, with significantly more *Il1β* in the *Asc*$^{-/-}$ BMDMs at 24 h compared to the wildtype (Figure 5B). Complementing the transcription data, we also evaluated the IL-1β protein levels in the cell supernatant using ELISA (Figure 5C). These levels were significantly attenuated in the *B. abortus*-infected cells (Figure 5C). We also observed a significant decrease in IL-1β in the *Asc*$^{-/-}$ BMDMs under all conditions (Figure 5C), emphasizing that IL-1β processing is ASC-dependent.

To further expand upon our cell death findings, we conducted a LDH assay in our *Asc*$^{-/-}$ and wildtype macrophages to quantify the lactate dehydrogenase enzyme released from dead cells (Figure 5D). We found that, starting at 24 h, there was significantly higher cell death occurring in our wildtype cells compared to the *Asc*$^{-/-}$ macrophages. To further define the mechanism of cell death, we evaluated pyroptosis by determining gasdermin D cleavage using Western blot. We found a significant increase in cleaved gasdermin D 24 h post-infection in the wildtype BMDMs compared to the significantly reduced levels observed in the *Asc*$^{-/-}$ macrophages (Figure 5E). This was confirmed using densitometry (Figure 5E).

2.6. *B. abortus* gDNA is a Potent PAMP Associated with ASC-Dependent Canonical Inflammasome Signaling

To further define inflammasome activation following *B. abortus* infection, we next evaluated the potential pathogen-associated molecular patterns (PAMPs), focusing on bacterial gDNA. We challenged macrophages with 1 μg of gDNA (2 μg/mL) both externally, by adding gDNA to the cell media, and internally through the Lipofectamine 3000 reagent (Figure 6A). We also added 300 μM of ATP to augment the IL-1β release. *B. abortus* gDNA induced a significant increase in the *Il1β* gene transcription 24 h post challenge, following either extracellular or intracellular challenge (Figure 6A). We observed significant increases in *Il1β* transcription under several different conditions in gDNA-challenged *Asc*$^{-/-}$ macrophages compared to the wildtype cells. ELISA assessments revealed that IL-1β protein was only released into the supernatant following internal gDNA stimulation in wildtype cells (Figure 6B). This was highly dependent on ASC. The *B. abortus* gDNA challenge resulted in significant increases in IL-1β in the wildtype cells, whereas the levels were below the level of detection in the *Asc*$^{-/-}$ cells (Figure 6B). Together, these data confirm *B. abortus* gDNA as a potent PAMP and suggest that its recognition by the canonical inflammasome, in an ASC-dependent mechanism, underlies host defense. In addition to gDNA, we also evaluated the ability of the canonical inflammasome to recognize *B. abortus* LPS (1 µg/mL externally). The fold change in *Il1β* RNA at 8 h post challenge indicated that there was elevated transcription in wildtype macrophages over $Asc^{-/-}$ cells (Figure 6C). However, there was no IL-1β protein signaling for *Brucella* LPS (Figure 6D).

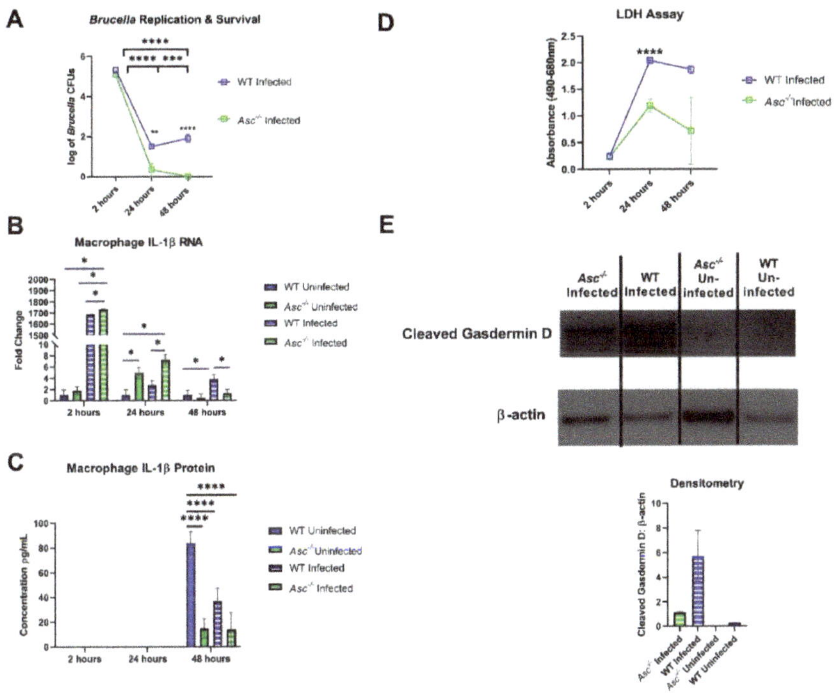

Figure 5. ASC-dependent IL-1β generation and pyroptosis following *B. abortus* infection in BMDMs. Bone marrow-derived macrophages (BMDMs) (500,000 cells per well, $n = 3$ per group) were harvested from C57BL/6 WT and $Asc^{-/-}$ mice and challenged with a MOI 100:1 (10^7 CFUs) of *Brucella abortus*. (**A**) BMDMs were measured for CFUs for a 48 h period post-challenge. (**B**) BMDMs were measured for the *Il1β* fold change in RNA through RT-PCR. (**C**) BMDMs were analyzed for IL-1β protein in the supernatant using an ELISA. (**D**) Lactate Dehydrogenase (LDH) was measured through spectrophotometry through a 48 h time period post-challenge. (**E**) BMDMs were lysed at 24 h post-challenge and used for Western blot analysis. BMDM protein (20µg) was probed for cleaved gasdermin D and β-actin as a control for the protein amount. Invitrogen IBright Analysis was used to determine the density ratio between the cleaved gasdermin D over β-actin. * $p < 0.05$, *** $p < 0.001$, **** $p < 0.0001$.

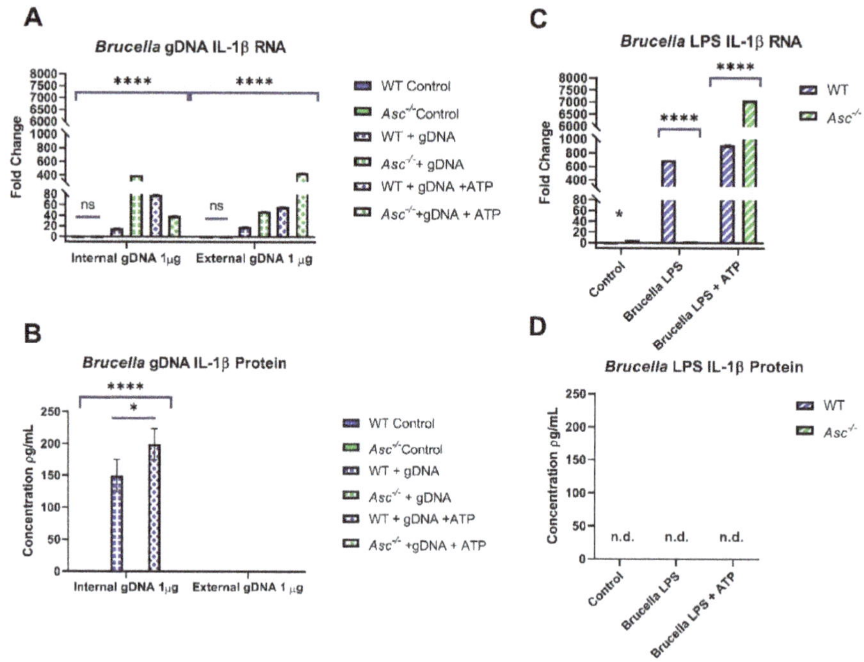

Figure 6. *B. abortus* gDNA is a potent PAMP and induces ASC-dependent IL-1β production. Bone marrow-derived macrophages (BMDMs) (500,000 cells per well, $n = 2$ per group) were harvested from C57BL/6 WT and *Asc*$^{-/-}$ mice. (**A**,**B**) BMDMs were challenged with 1 μg of gDNA (2μg/mL) externally in media or transfected internally with 300 μM of ATP and harvested after 24 h. BMDMs were analyzed for the IL-1β (**A**) RNA fold change through RT-PCR and (**B**) protein concentration through ELISA. (**C**,**D**) BMDMs were challenged with 1 μg/mL of *Brucella* LPS externally to the media with 300 μM of ATP and harvested after 8 h. BMDMs were analyzed for the IL-1β (**C**) RNA fold change through RT-PCR and (**D**) protein concentration through ELISA. Note that there was no detectable (*n*.d.) protein concentration in (**D**). * $p < 0.05$, **** $p < 0.0001$.

3. Discussion

In this report, we assessed inflammasome activation following *Brucella abortus* infection. During canonical inflammasome activation, a pathogen is sensed by a NLR or ALR pattern recognition receptor and forms a multi-protein complex with the binding protein ASC and caspase-1. This process can initiate the cleavage of IL-1β and IL-18 in addition to pyroptosis [8]. While highly related to the canonical inflammasome pathway, non-canonical inflammasome activation is often more closely associated with the promotion of pyroptosis and the activation of caspase-11 [26]. Together, our in vivo data reveal a more robust phenotype in the *Asc*$^{-/-}$ mice compared to the *Caspase-11*$^{-/-}$ animals, suggesting that the canonical inflammasome plays a significantly greater role in host pathogen defense following *B. abortus* infection. ASC and, by extension, the canonical inflammasome promotes survival, augments inflammation, and attenuates bacterial load *in vivo*. This result is consistent with others in the field that have used myriad of other inflammasome knockout models [5–7,27]. However, there are certainly exceptions to these findings. For example, Gomes et al. (2103) recently reported no inflammasome knockout mice ($n = 5$ per group) exhibited mortality under their experimental conditions [6]. One difference to note is that our study utilized a much larger sample size per group ($n = 19$). Because many of our animals recovered, it is certainly possible that greater power through that increased sample size is necessary to better reflect the mortality data for the *Asc*$^{-/-}$ animals. Ultimately, our comparison between *Asc*$^{-/-}$ and *Caspase-11*$^{-/-}$ mice expands upon the findings of many of these prior studies and provides a more direct assessment of the pathobiological effects of the canonical and non-canonical inflammasome in *B. abortus* host defense.

Although non-canonical signaling through the *Brucella* LPS activation of caspase-11-mediated pyroptosis has been previously indicated [28], we found inconsistent results in inflammasome activation under our conditions. In *Caspase-11*$^{-/-}$ mice, we found no loss in survival or morbidity, no significant inflammation through histopathology scoring, and no protective role in promoting inflammatory signaling. Previous studies using *Caspase-11*$^{-/-}$ mice mimicked our non-significant results in bacterial load at 3 days post-infection and only found significant bacterial load 1–2 weeks after infection [28]. Although we did not assess the bacterial load or inflammatory signaling after 3 d.p.i in this report, we did assess the morbidity and mortality of *Caspase-11*$^{-/-}$ mice over 3 weeks, in which knockouts exhibited no decrease in morbidity or loss in survival. This indicates that caspase-11 may play a small but relatively insignificant role in promoting pyroptosis that perhaps may be slightly amplified 1–2 weeks after *Brucella* infection. Additionally, these studies utilized immune cell priming in macrophages with PAMPs, such as *E. coli* LPS and Pam3CSK4, and subsequently observed high cytokine signaling in their challenge [28]. Contrasting this data, our results indicated no IL-1β protein response to *Brucella* LPS stimulation in unprimed macrophages. These data suggest that the immune adjuvants directly impact the activation of caspase-11 and that the role of non-canonical inflammasome in host defense against *B. abortus* is minimal in the absence of macrophage priming.

Previously, pyroptosis had only been attributed to non-canonical inflammasome activation by *Brucella* spp. [6,28]. This is further confirmed in this report through the presence of cleaved gasdermin D bands in *Asc*$^{-/-}$ macrophages, indicating that ASC-independent pyroptosis occurs in response to *Brucella*. However, our research demonstrates that the removal of ASC-dependent inflammasome activation significantly decreases the activation of gasdermin D to form pyroptotic pores. As described in the literature, ASC specks serve as recruitment factors for procaspase-1 through the polymerization of its caspase activation recruitment domain (CARD). Caspase-1 only becomes activated through this process during ASC-dependent inflammasome formation [29]. In turn, caspase-1 cleaves gasdermin D, which has been identified as the most significant gene initiating caspase-1 induced pyroptosis, and initiates inflammatory cell death [30]. Our results are consistent with this ASC-mediated pathway of the capsase-1 activation of pyroptosis that is dependent on gasdermin D cleavage. To date, we know that *Brucella* initiates the caspase-1 and -11 activation of pyroptosis in joints of animal models [31]. Additionally, pyroptosis is activated by gDNA in dendritic cells [5]. Our data supports a model where both caspase-1 and -11 promote pyroptosis, and where gDNA from *B. abortus* functions as a robust PAMP that specifically activates the canonical inflammasome, driving ASC-dependent inflammation and pyroptosis.

Previous literature indicates that the role of pyroptosis in brucellosis serves to restrict *Brucella* growth in macrophages of the joints and control infection [31]. Our findings are consistent with this previous study. Under our conditions, the ASC-mediated initiation of pyroptosis appears to ensure mouse survival, immune cell recruitment, and inflammatory signaling. However, we should also point out the bacteria clearance and IL-1β data in the BMDM studies (Figure 5A,C). In Figure 5A, these data would suggest that the lack of ASC and canonical inflammasome signaling actually improved the bacteria clearance from these BMDMs, despite having reduced IL-1β and pyroptosis. While these data seem to conflict with each other, several recent studies have reported similar findings for other bacterial pathogens. For example, *Citrobacter rodentium* infection results in significant osmotic changes in targeted cells that can augment inflammasome signaling [32]. However, the clearance of the pathogen itself appears to be independent of the inflammasome and the ASC modulation of inflammation and pyroptosis [33]. Thus, it is possible that a similar mechanism is associated here with *B. abortus*. Looking at the IL-1β graph in Figure 5C, it suggests that *B. abortus* infection in macrophages suppresses IL-1β production from wildtype cells. It is possible that the attenuation of total IL-1β in this figure may be due to the subversion of TLR signaling generating pro- IL-1β. Many studies have shown *Brucella* subversion of TLR signaling through proteins such as Tcbp, which leads to decreased proinflammatory cytokine expression [10]. Additionally, *Brucella* microRNAs can lead to the downregulation of the mRNA and protein expression of innate immune PRRs [34]. These mechanisms likely did not occur in our gDNA studies as there was no inclusion of these immunosuppressive proteins or production of inhibitory microRNAs. Therefore, it is possible that the full *Brucella* bacterium may be utilizing these methods of immunosuppression to contribute to decreased total IL-1β.

Our data suggests that ASC and the canonical inflammasome contribute to host defense in response to *B. abortus*. These results are consistent with several others in the field and provide additional insight into host defense against this highly intriguing pathogen. Currently, brucellosis is having a significantly negative impact on a growing number of human populations worldwide. Therefore, it is essential that we expand our understanding of the underlying disease mechanisms and host immune response to *B. abortus*. This finding that the canonical inflammasome plays a dominate role in driving the host innate immune response and pyroptosis following the sensing of gDNA is an encouraging discovery that may contribute to the development of future therapeutics or strategic approaches to combat this disease and its underlying pathogen.

4. Materials and Methods

4.1. Bacterial Strains and Growth Conditions

Brucella abortus 2308 was routinely grown on Schaedler blood agar (SBA), which is composed of Schaedler agar (BD, Franklin Lakes, NJ, USA) containing 5% defibrinated bovine blood (Quad Five, Ryegate, MT, USA). All work with live *Brucella* strains was performed in a biosafety level 3 (BSL3) facility. Animal work conducted in ABSL3 conditions was conducted **under IACUC protocol # 14-055** at Virginia Tech following the ethical standards of animal use in research.

4.2. In Vivo Brucella Studies

C57BL/6 WT mice (n = 26), *Asc*$^{-/-}$ mice (C57BL/6 background; n = 19), and *Caspase-11*$^{-/-}$ (C57BL/6 background; n = 19) mice were inoculated intraperitoneally with 1×10^5 CFU of *Brucella abortus*. The percent weight change was measured each day post-*Brucella* inoculation to determine morbidity warranting euthanasia (>20%) following the IACUC protocol.

Additionally, 20 canonical knockout (*Asc*$^{-/-}$, n = 6 U, 14 I), 27 non-canonical (*Caspase-11*$^{-/-}$, n = 8 U, 19 I), and 34 C57BL/6 WT (n = 11 U, 23 I) mice were euthanized at 3 days post infection and harvested for the liver and spleen. Both organs were sectioned into three equal parts. The first section of both organs was taken for histopathology analysis to determine the scoring of extramedullary hematopoiesis (EMH) and inflammation. The remaining sections were homogenized in 1×PBS and analyzed for the number of CFUs per gram and the RNA and protein concentrations of IL-1β. RNA was isolated from the liver and spleen with TRIzol reagents (Invitrogen) followed by ethanol precipitation. Genomic DNA was removed with DNase I, and samples were cleaned using phenol-chloroform extractions and precipitated with ethanol. RNA samples were then resuspended in nuclease-free H_2O, and the purity of samples was checked with a NanoDrop 1000 spectrophotometer (ThermoFisher).

After isolation, RNA was converted into 1 μg cDNA through a High-Capacity cDNA Reverse Transcription Kit (ThermoFisher). This cDNA was analyzed for IL-1β by RT-qPCR (40× cycles) using Taqman Fast MasterMix (ThermoFisher). Protein was determined through a sandwich ELISA kit (R&D systems).

4.3. Bone Marrow-Derived Macrophage (BMDM) Isolation

Bone marrow-derived macrophages (BMDMs) were derived from the bone marrow of C57BL/6 WT and *Asc*$^{-/-}$ mice under the **IACUC protocol #18-104** at Virginia Tech. Two adult mice from each group were sacrificed using CO_2 fixation followed by cervical dislocation. Bone marrow was extracted from the tibias and femora of the mice. Cells were cultured in non-TC culture dishes in our formulation of culture media (Dulbecco's Modified Eagle Medium (DMEM) (ThermoFisher) containing 10% Fetal Bovine Serum (FBS), 1% penicillin/streptomycin, 1% nonessential amino acids, and 20% L929 conditioned media [35]). These culture dishes were incubated at 37 °C with 5% CO_2. After 6–7 days of culture, these cells had differentiated into macrophages. BMDMs were collected from the bottom of the plates through a cold 1× PBS solution containing 5 mM of EDTA. After 1 h on ice, we checked for macrophage detachment via microscope. Macrophages were collected and seeded at 500,000 cells/well in 24-well plates in media without antibiotics and left to adhere overnight in culture media.

4.4. *Live Brucella Challenge in BMDMs*

Macrophages were infected with a MOI 100:1 (10^7 CFUs/5 × 10^5 BMDMs) with *B. abortus* 2308. At the 2, 24, and 48 h time points, intracellular *B. abortus* was determined through a gentamicin protection assay [36]. At 2 h, the infected macrophages were treated with gentamicin (50 µg/mL) for 1 h. Macrophages were then lysed with 0.1% deoxycholate in PBS, and serial dilutions were plated on Schaedler blood agar (SBA) containing 5% bovine blood (Quad Five). For the 24 and 48 h time points, macrophages were washed with PBS and fresh cell culture medium containing gentamicin (20 µg/mL) was added to the macrophages. At the indicated time points, macrophages were lysed, and serial dilutions were plated on SBA in triplicates.

The supernatant of BMDMs was sterile-filtered for later IL-1β protein analysis through a sandwich ELISA kit (R&D systems). Macrophages at each time point were lysed with 0.1% deoxycholate and isolated for RNA. After isolation, RNA was converted into 0.5µg cDNA through a High-Capacity cDNA Reverse Transcription Kit (ThermoFisher). This cDNA was analyzed for IL-1β by RT-qPCR (40× cycles) using Taqman Fast MasterMix (ThermoFisher).

4.5. *LDH Assay*

Macrophages were infected with a MOI 100:1 (10^7 CFUs/5 × 10^5 BMDMs) with *B. abortus*. At the 2, 24, and 48 h time points, extracellular *B. abortus* was killed through gentamicin (50 µg/mL). The supernatant of BMDMs was collected and centrifuged at 1000× *g* for 10 min to remove cell debris. The remaining supernatant (50 µL) was used for the CyQUANT™ LDH Cytotoxicity Assay kit (Invitrogen) and read at a corrected absorbance of 490–680 nm.

4.6. *Brucella PAMP Isolation*

4.6.1. *Brucella* gDNA

B. abortus gDNA was isolated from the 2308 strain by phenol:chloroform extraction. Approximately 3 mL of an overnight culture of *B. abortus* was pelleted by centrifugation. The pellet was resuspended in 200 µL of 0.04 M sodium acetate, 200 µL of 10% SDS, and 600 µL of TRIzol. A total of 250 µL of chloroform was added to this mixture in a Phase Lock tube (5PRIME) and centrifuged at 20,000× *g* (max speed) for 2 min. A second chloroform wash was performed in a Phase Lock tube. The resulting aqueous layer is then removed and added to 1 mL of 100% ethanol. DNA precipitation is carried out from this point.

4.6.2. *Brucella* LPS

B. abortus LPS was isolated from the *B. abortus* 2308 using hot-phenol extraction, as described previously [31]. Bacteria were killed with ethanol: acetone, and the cells were recovered by centrifugation. The pellet was suspended in deionized water at 66 °C and then mixed with 90% phenol *w/v* that was heated to 66 °C. After stirring for 20 min, the suspension was chilled on ice. The solution was then subjected to centrifugation (15 min at 13,000× *g*). The phenol layer was aspirated, filtered through a Whatman #1 filter, and the LPS was precipitated with methanol containing 1% methanol saturated with sodium acetate. Following incubation at 4 °C for 1 h, the mixture was subjected to centrifugation (10,000× *g* for 10 min). The precipitate was stirred with deionized water for 12 h at 4 °C.

Following centrifugation (10,000× *g* for 10 min), the supernatant was precipitated with trichloroacetic acid, and the resulting supernatant following centrifugation (10,000× *g* for 10 min) was dialyzed with deionized water and stored at −20 °C. The concentration of LPS was determined through the Pierce™ Chromogenic Endotoxin Quant Kit (ThermoFisher).

4.7. *Brucella PAMP Challenge*

After PAMP isolation, *B. abortus* gDNA (2 μg/mL) and LPS (1 μg/mL) were introduced to BMDMs. gDNA was introduced both intracellularly, through the Lipofectamine 3000 Transfection Reagent (Invitrogen), and extracellularly in media. LPS was only introduced extracellularly. Timepoints for this challenge included 24 h for gDNA and 8 h for LPS post-challenge. Samples were run with and without the priming of 300 μM of ATP 45 min before each time point to stimulate IL-1β protein release after transcription. At each time point, supernatant was collected from each well and centrifuged at 1000× *g* for 10 min to remove cell debris. The supernatant was later used for protein quantification using an IL-1β sandwich ELISA (R&D Systems). Macrophages were lysed with 200 μL of TRIzol Reagent (Invitrogen) and followed the TRIzol RNA isolation protocol. After isolation, RNA was converted into 1 μg of cDNA through a High-Capacity cDNA Reverse Transcription Kit (ThermoFisher). This cDNA was analyzed for IL-1β by RT-qPCR (40× cycles) using the Taqman Fast MasterMix (ThermoFisher).

4.8. *Western Blot*

Macrophages from live *Brucella* challenge were lysed using a sodium lysis buffer (0.3% SDS, 200 mM dithiothreitol, 22 mM Tris-base, and 28 mM Tris-HCl pH 8.0). Samples were then boiled for a period of 1 h, vortexing the samples every 10 min. Samples were frozen at −20 °C until use. Prior to running the gel, samples were sonicated for 10 s each. Protein quantification was determined using the Pierce™ Detergent Compatible Bradford Assay Kit (ThermoFisher). Samples (20 μg of protein) were heated at 97 °C with a reducing buffer for 7 min and run on pre-cast Bolt™ 4 to 12%, Bis-Tris, 1.0 mm, Mini Protein Gel, 10-well gels (Invitrogen) for 45 min at 165 V with a 1× Micro Extraction Packet Sorbent (MEPS) buffer (ThermoFisher). Gel was transferred onto a polyvinylidene difluoride (PVDF) membrane using a transfer chamber with transfer buffer (20% methanol in 1× Tris Glycine (TGE)). The membrane was then blocked in 5% milk for 1 h, and then incubated overnight with a cleaved gasdermin D rabbit antibody (diluted 1:1000, Cell Signaling #36425S). Membrane was washed in Tris-buffered saline with Tween 20 (1× TBST) 4× for 15 min each and then blocked with goat anti-rabbit IgG antibody conjugated with horseradish peroxidase (HRP) (diluted 1:2000, Cell Signaling #7074) for 1 h. Membrane was washed again with 1× TBST 4× for 15 min each and then imaged using West Pico Substrate for imaging (ThermoFisher). Gels were imaged using an IBright CL1500 imaging machine. Sample bands were normalized using the β-actin rabbit antibody (Cell Signaling #4970) using the same protocol above. The density of bands was calculated using the IBright Analysis software (ThermoFisher) to calculate a cleaved gasdermin D/β-actin ratio.

4.9. *Graphing and Statistical Analyses*

All the figures and statistical analyses were generated in GraphPad 8.4.3 (Prism). Statistical tests included two-way ANOVAs, using the Tukey or Sidak post-hoc tests, and two sample *t*-tests when appropriate. All the data are contained within the article.

Author Contributions: The authors contributed to the following project aspects: Conceptualization: I.C.A., C.C.C., J.D.T.; methodology: all authors; software: J.D.T.; validation: all authors; formal analysis: J.D.T. and S.L.C.-O.; investigation: J.D.T., S.L.C.-O., A.H.B., K.A.K., H.D.K.; resources: C.C.C. and I.C.A.; data curation: J.D.T. and S.L.C.-O.; writing—original draft preparation: all authors; writing—review and editing: all authors; visualization: J.D.T.; supervision: C.C.C. and I.C.A.; project administration: C.C.C. and I.C.A.; funding acquisition: C.C.C. and I.C.A. All authors have read and agreed to the published version of the manuscript.

Funding: This research was funded by the Virginia Maryland College of Veterinary Medicine (VT/UMD Joint Seed Grant; DVM/PhD Student Support) and the NIH/NIAID (R03AI151494)

Acknowledgments: We would like to acknowledge the assistance of Margaret Nagai-Singer, Alissa Hendricks-Wenger, Jackie Sereno, and Audrey Rowe with conducting methods. Additionally, we are grateful to Jerod Skyberg for providing detailed information about LPS isolation from *Brucella* strains.

Conflicts of Interest: The authors declare no conflict of interest.

References

1. Corbel, M.J. Brucellosis: An overview. *Emerg. Infect. Dis.* **1997**, *3*, 213. [CrossRef] [PubMed]
2. Lapaque, N.; Moriyon, I.; Moreno, E.; Gorvel, J.P. *Brucella* lipopolysaccharide acts as a virulence factor. *Curr. Opin. Microbiol.* **2005**, *8*, 60–66. [CrossRef] [PubMed]
3. De Figueiredo, P.; Ficht, T.A.; Rice-Ficht, A.; Rossetti, C.A.; Adams, L.G. Pathogenesis and immunobiology of brucellosis: Review of *Brucella*–Host Interactions. *Am. J. Pathol.* **2015**, *185*, 1505–1517. [CrossRef] [PubMed]
4. Janeway, C.A. Approaching the asymptote? Evolution and revolution in immunology. *Cold Spring Harb. Symp. Quant. Biol.* **1989**, *54*, 1–13. [CrossRef] [PubMed]
5. Franco, M.M.S.C.; Marim, F.M.; Alves-Silva, J.; Cerqueira, D.; Rungue, M.; Tavares, I.P.; Oliveira, S.C. AIM2 senses *Brucella abortus* DNA in dendritic cells to induce IL-1β secretion, pyroptosis and resistance to bacterial infection in mice. *Microbes Infect.* **2019**, *21*, 85–93. [CrossRef]
6. Gomes, M.T.R.; Campos, P.C.; Oliveira, F.S.; Corsetti, P.P.; Bortoluci, K.R.; Cunha, L.D.; Zamboni, D.S.; Oliveira, S.C. Critical role of ASC inflammasomes and bacterial type IV secretion system in caspase-1 activation and host innate resistance to *Brucella abortus* infection. *J. Immunol.* **2013**, *190*, 3629–3638. [CrossRef]
7. Marim, F.M.; Franco, M.M.C.; Gomes, M.T.R.; Miraglia, M.C.; Giambartolomei, G.H.; Oliveira, S.C. The role of NLRP3 and AIM2 in inflammasome activation during *Brucella abortus* infection. *Semin. Immunopathol.* **2017**, *39*, 215–223. [CrossRef]
8. Coutermarsh-Ott, S.; Eden, K.; Allen, I.C. Beyond the inflammasome: Regulatory NOD-like receptor modulation of the host immune response following virus exposure. *J. Gen. Virol.* **2016**, *97*, 825–838. [CrossRef]
9. Tupik, J.D.; Nagai-Singer, M.A.; Allen, I.C. To protect or adversely affect? The dichotomous role of the NLRP1 inflammasome in human disease. *Mol. Asp. Med.* **2020**, 100858. [CrossRef]
10. Oliveira, S.C.; de Oliveira, F.S.; Macedo, G.C.; de Almeida, L.A.; Carvalho, N.B. The role of innate immune receptors in the control of *Brucella abortus* infection: Toll-like receptors and beyond. *Microbes Infect.* **2008**, *10*, 1005–1009. [CrossRef]
11. Campos, M.A.; Rosinha, G.M.; Almeida, I.C.; Salgueiro, X.S.; Jarvis, B.W.; Splitter, G.A.; Qureshi, N.; Bruna-Romero, O.; Gazzinelli, R.T.; Oliveira, S.C. Role of Toll-like receptor 4 in induction of cell-mediated immunity and resistance to *Brucella abortus* infection in mice. *Infect. Immun.* **2004**, *72*, 176–186. [CrossRef] [PubMed]
12. De Almeida, L.A.; Macedo, G.C.; Marinho, F.A.; Gomes, M.T.; Corsetti, P.P.; Silva, A.M.; Cassataro, J.; Giambartolomei, G.H.; Oliveira, S.C. Toll-like receptor 6 plays an important role in host innate resistance to *Brucella abortus* infection in mice. *Infect. Immun.* **2013**, *81*, 1654–1662. [CrossRef] [PubMed]
13. Ferrero, M.C.; Hielpos, M.S.; Carvalho, N.B.; Barrionuevo, P.; Corsetti, P.P.; Giambartolomei, G.H.; Oliveira, S.C.; Baldi, P.C. Key role of Toll-like receptor 2 in the inflammatory response and major histocompatibility complex class II downregulation in *Brucella abortus*-infected alveolar macrophages. *Infect. Immun.* **2014**, *82*, 626–639. [CrossRef] [PubMed]
14. Campos, P.C.; Gomes, M.T.R.; Guimarães, E.S.; Guimarães, G.; Oliveira, S.C. TLR7 and TLR3 sense *Brucella abortus* RNA to induce proinflammatory cytokine production but they are dispensable for host control of infection. *Front. Immunol.* **2017**, *8*, 28. [CrossRef] [PubMed]
15. Macedo, G.C.; Magnani, D.M.; Carvalho, N.B.; Bruna-Romero, O.; Gazzinelli, R.T.; Oliveira, S.C. Central role of MyD88-dependent dendritic cell maturation and proinflammatory cytokine production to control *Brucella abortus* infection. *J. Immunol.* **2008**, *180*, 1080–1087. [CrossRef] [PubMed]
16. De Almeida, L.A.; Carvalho, N.B.; Oliveira, F.S.; Lacerda, T.L.S.; Vasconcelos, A.C.; Nogueira, L.; Bafica, A.; Silva, A.M.; Oliveira, S.C. MyD88 and STING signaling pathways are required for IRF3-mediated IFN-β induction in response to *Brucella abortus* infection. *PLoS ONE* **2011**, *6*, e23135. [CrossRef]
17. Hornung, V.; Latz, E. Critical functions of priming and lysosomal damage for NLRP3 activation. *Eur. J. Immunol.* **2010**, *40*, 620–623. [CrossRef]

18. Miao, E.A.; Andersen-Nissen, E.; Warren, S.E.; Aderem, A. TLR5 and Ipaf: Dual sensors of bacterial flagellin in the innate immune system. *Semin. Immunopathol.* **2007**, *29*, 275–288. [CrossRef]
19. Martinon, F.; Burns, K.; Tschopp, J. The inflammasome: A molecular platform triggering activation of inflammatory caspases and processing of proIL-beta. *Mol. Cell* **2002**, *10*, 417–426. [CrossRef]
20. Agostini, L.; Martinon, F.; Burns, K.; McDermott, M.F.; Hawkins, P.N.; Tschopp, J. NALP3 forms an IL-1beta-processing inflammasome with increased activity in Muckle-Wells autoinflammatory disorder. *Immunity* **2004**, *20*, 319–325. [CrossRef]
21. Davis, B.K.; Roberts, R.A.; Huang, M.T.; Willingham, S.B.; Conti, B.J.; Brickey, W.J.; Barker, B.R.; Kwan, M.; Taxman, D.J.; Accavitti-Loper, M.A.; et al. Cutting edge: NLRC5-dependent activation of the inflammasome. *J. Immunol.* **2011**, *186*, 1333–1337. [CrossRef] [PubMed]
22. Wlodarska, M.; Thaiss, C.A.; Nowarski, R.; Henao-Mejia, J.; Zhang, J.P.; Brown, E.M.; Frankel, G.; Levy, M.; Katz, M.N.; Philbrick, W.M.; et al. NLRP6 inflammasome orchestrates the colonic host-microbial interface by regulating goblet cell mucus secretion. *Cell* **2014**, *156*, 1045–1059. [CrossRef] [PubMed]
23. Vanaja, S.K.; Rathinam, V.A.K.; Fitzgerald, K.A. Mechanisms of inflammasome activation: Recent advances and novel insights. *Trends Cell Biol.* **2015**, *25*, 308–315. [CrossRef] [PubMed]
24. Shi, J.; Gao, W.; Shao, F. Pyroptosis: Gasdermin-mediated programmed necrotic cell death. *Trends Biochem. Sci.* **2017**, *42*, 245–254. [CrossRef]
25. Franco, M.M.C.; Marim, F.; Guimarães, E.S.; Assis, N.R.; Cerqueira, D.M.; Alves-Silva, J.; Harms, J.; Splitter, G.; Smith, J.; Kanneganti, T.-D. *Brucella abortus* triggers a cGAS-independent STING pathway to induce host protection that involves guanylate-binding proteins and inflammasome activation. *J. Immunol.* **2018**, *200*, 607–622. [CrossRef] [PubMed]
26. Yi, Y.-S. Regulatory Roles of the Caspase-11 Non-Canonical Inflammasome in Inflammatory Diseases. *Immune Netw.* **2018**, *18*, e41. [CrossRef]
27. Hielpos, M.S.; Fernández, A.G.; Falivene, J.; Alonso Paiva, I.M.; Muñoz González, F.; Ferrero, M.C.; Campos, P.C.; Vieira, A.T.; Oliveira, S.C.; Baldi, P.C. IL-1R and inflammasomes mediate early pulmonary protective mechanisms in respiratory *Brucella abortus* infection. *Front. Cell. Infect. Microbiol.* **2018**, *8*, 391. [CrossRef]
28. Cerqueira, D.M.; Gomes, M.T.R.; Silva, A.L.; Rungue, M.; Assis, N.R.; Guimarães, E.S.; Morais, S.B.; Broz, P.; Zamboni, D.S.; Oliveira, S.C. Guanylate-binding protein 5 licenses caspase-11 for Gasdermin-D mediated host resistance to *Brucella abortus* infection. *PLoS Pathog.* **2018**, *14*, e1007519. [CrossRef]
29. Lu, A.; Magupalli, V.G.; Ruan, J.; Yin, Q.; Atianand, M.K.; Vos, M.R.; Schröder, G.F.; Fitzgerald, K.A.; Wu, H.; Egelman, E.H. Unified polymerization mechanism for the assembly of ASC-dependent inflammasomes. *Cell* **2014**, *156*, 1193–1206. [CrossRef]
30. Broz, P. Caspase target drives pyroptosis. *Nature* **2015**, *526*, 642–643. [CrossRef]
31. Lacey, C.A.; Mitchell, W.J.; Dadelahi, A.S.; Skyberg, J.A. Caspase-1 and caspase-11 mediate pyroptosis, inflammation, and control of *Brucella* joint infection. *Infect. Immun.* **2018**, *86*. [CrossRef]
32. Liu, Z.; Zaki, M.H.; Vogel, P.; Gurung, P.; Finlay, B.B.; Deng, W.; Lamkanfi, M.; Kanneganti, T.D. Role of inflammasomes in host defense against *Citrobacter rodentium* infection. *J. Biol. Chem.* **2012**, *287*, 16955–16964. [CrossRef] [PubMed]
33. Armstrong, H.; Bording-Jorgensen, M.; Chan, R.; Wine, E. Nigericin Promotes NLRP3-Independent Bacterial Killing in Macrophages. *Front. Immunol.* **2019**, *10*, 2296. [CrossRef] [PubMed]
34. Khan, M.; Harms, J.S.; Liu, Y.; Eickhoff, J.; Tan, J.W.; Hu, T.; Cai, F.; Guimaraes, E.; Oliveira, S.C.; Dahl, R.; et al. *Brucella* suppress STING expression via miR-24 to enhance infection. *PLoS Pathog.* **2020**, *16*, e1009020. [CrossRef] [PubMed]

35. Englen, M.D.; Valdez, Y.E.; Lehnert, N.M.; Lehnert, B.E. Granulocyte/macrophage colony-stimulating factor is expressed and secreted in cultures of murine L929 cells. *J. Immunol. Methods* **1995**, *184*, 281–283. [CrossRef]
36. Sharma, A.; Puhar, A. Gentamicin Protection Assay to Determine the Number of Intracellular Bacteria during Infection of Human TC7 Intestinal Epithelial Cells by Shigella flexneri. *Bio-Protocol* **2019**, *9*. [CrossRef]

Publisher's Note: MDPI stays neutral with regard to jurisdictional claims in published maps and institutional affiliations.

pathogens MDPI

Review

Brucella: Reservoirs and Niches in Animals and Humans

Gabriela González-Espinoza, Vilma Arce-Gorvel, Sylvie Mémet *,† and Jean-Pierre Gorvel *,†

Aix Marseille Univ, CNRS, INSERM, CIML, Centre d'Immunologie de Marseille-Luminy, 13288 Marseille, France; gonzalez@ciml.univ-mrs.fr (G.G.-E.); arce@ciml.univ-mrs.fr (V.A.-G.)
* Correspondence: memet@ciml.univ-mrs.fr (S.M.); gorvel@ciml.univ-mrs.fr (J.-P.G.)
† Co-senior authorship.

Abstract: *Brucella* is an intracellular bacterium that causes abortion, reproduction failure in livestock and leads to a debilitating flu-like illness with serious chronic complications if untreated in humans. As a successful intracellular pathogen, *Brucella* has developed strategies to avoid recognition by the immune system of the host and promote its survival and replication. In vivo, Brucellae reside mostly within phagocytes and other cells including trophoblasts, where they establish a preferred replicative niche inside the endoplasmic reticulum. This process is central as it gives *Brucella* the ability to maintain replicating-surviving cycles for long periods of time, even at low bacterial numbers, in its cellular niches. In this review, we propose that *Brucella* takes advantage of the environment provided by the cellular niches in which it resides to generate reservoirs and disseminate to other organs. We will discuss how the favored cellular niches for *Brucella* infection in the host give rise to anatomical reservoirs that may lead to chronic infections or persistence in asymptomatic subjects, and which may be considered as a threat for further contamination. A special emphasis will be put on bone marrow, lymph nodes, reproductive and for the first time adipose tissues, as well as wildlife reservoirs.

Keywords: *Brucella*; replication niche; reservoir; persistence; survival; chronic infection

Citation: González-Espinoza, G.; Arce-Gorvel, V.; Mémet, S.; Gorvel, J.-P. *Brucella*: Reservoirs and Niches in Animals and Humans. *Pathogens* **2021**, *10*, 186. https://doi.org/10.3390/pathogens10020186

Academic Editor: Sergio Costa Oliveira
Received: 14 January 2021
Accepted: 5 February 2021
Published: 9 February 2021

1. Introduction

Brucella is a Gram-negative facultative intracellular coccobacillus, responsible for brucellosis, a worldwide zoonotic disease that affects livestock, wildlife, and humans. Brucellosis remains endemic in many parts of the world, including the Middle East, Africa, Latin America, Central Asia, and several regions of the Mediterranean [1–4]. *Brucella melitensis*, *Brucella abortus*, and *Brucella suis* are the most important members of the genus because they are responsible for the human disease [5,6] and for significant economic losses in livestock [7,8], while *Brucella canis*, *Brucella neotomae*, and *Brucella ovis* display less or non-zoonotic potential [9]. Recently, many other species have been described in aquatic mammals (like *Brucella pinnipedialis* and *Brucella ceti*) and other wildlife such as *Brucella papionis*, *Brucella microti*, *Brucella inopinata*, and *Brucella vulpis* [10–16].

1.1. Brucellosis, a Zoonosis

Humans are accidental hosts for *Brucella* and are mainly infected through direct contact with infected animals, inhalation of airborne agents, or by ingestion of contaminated dairy products [17]. Human-to-human transmission may occur by organ transplantation, blood transfusion, or vertical transmission via breastfeeding [18–22]. Human brucellosis is at the origin of many symptoms namely undulating fever, malaise, fatigue, and anorexia. If untreated, it may progress into a chronic phase, characterized by the appearance of severe complications like endocarditis, orchitis, spondylitis, osteomyelitis, arthritis, meningoencephalitis, and recurring febrile conditions [23–26].

In domestic animals, such as cattle, sheep, goats, and swine, major consequences include abortion and metritis in females, and orchiepididymitis and infertility in males [27], resulting in reduced fertility and a significant decline in milk production [28]. Animal brucellosis is highly contagious for both animals and humans, and cross-species transmission of certain *Brucella* spp. exists [29]. Natural infection occurs by direct contact with infected animals or their secretions [30], like aborted fetuses and fetal membranes that contain large amounts of the bacteria [31].

1.2. Infection and Dissemination

Brucella organisms enter into their host through the mucosal membranes of the respiratory and digestive tracts [32]. Once inside, local professional phagocytes such as macrophages, dendritic cells, and neutrophils internalize the bacteria and move to the closest draining lymph nodes following the normal sampling of the immune system. This leads to subsequent dissemination to the different organs of the reticuloendothelial system, including lungs, spleen, liver, and bone marrow [33]. In pregnant animals, *Brucella* displays a strong tropism for placental trophoblasts [34–36] and also for mammary glands [37], in which it replicates extensively causing placentitis and abortion in the last trimester of pregnancy in ruminants [33]. In humans, brucellosis is a systemic infection and any organ can become infected, albeit with some predilection for joints and liver, and at lower levels for the brain and heart [38].

1.3. Acute and Chronic Infections

Human brucellosis presents a broad range of clinical manifestations from asymptomatic to severe disease. When symptoms are present, the incubation period is 1–4 weeks, but sometimes it takes several months with or without signs and symptoms. According to the duration of the symptoms, human brucellosis is classified into three different phases: acute (initial, 2 months), sub-acute (2–12 months), and chronic (more than 12 months) [39,40]. When focusing on the chronic phase, patients fall into three more categories: relapse (with fever or high IgG antibody titers after antibiotics therapy), chronic localized infection (recurrence of infectious foci and intermittent fever for long periods of time), and delayed convalescence (persistence of some symptoms without fever or other objective signs) [41].

In animals, brucellosis comprises three phases: the onset of the infection or incubation period, when *Brucella* invades the host without any clinical symptoms; the acute phase, when the bacteria replicate actively and infection remains unapparent in most cases or first pathological signs arise; and the chronic phase, during which bacterial loads reach a plateau before decreasing and sporadic clinical symptoms become visible, the infection localizing in sexually mature animals in the reproductive system to produce epididymitis or orchitis in males, and placentitis and abortion in females. Susceptibility to infection in females increases during the late pregnancy stages (third trimester of gestation) [31,42]. There is no pyrexia like in humans and death is rare; infection is self-limiting most of the time and does not involve other systemic lesions, but for *Brucella suis* infection in swine. Clinical symptoms for the latter include spondylitis of the lumbar and sacral regions, arthritis, paralysis of hind limbs, lameness, and abscesses in tissues [42]. In any case, chronicity relies on the continuous shedding of *Brucella* from the mammary gland or reproductive organ secretions for a protracted period. Therefore, infertility, repetitive abortions, and premature stillborn ensure *Brucella*'s permanence in the environment and liability for dissemination.

The long incubation period and the absence of obvious clinical signs in infected animals and patients correlate with low activation of innate immunity [43]. *Brucella* has indeed modified its pathogen-associated molecular patterns (PAMPs) into less detectable molecules, opening a permissive immunological window to spread stealthily throughout the reticuloendothelial system and find its target host cells [44]. Then, invasion of target cells occurs thanks to specific molecular determinants that drive not only ingress [45] but also resistance to intracellular killing allowing *Brucella* to reach its intracellular replicative niche within professional and non-professional phagocytes [46]. Even if innate immunity first restrains *Brucella* proliferation, a Th1 response with IFNγ and IL-12 production is absolutely required to eradicate it, leading either to complete clearance of the bacteria or to a chronic infection, which arises from the ability of *Brucella* to persist undetected for prolonged periods of time within its host reservoirs [44,47].

We propose that the intracellular replication niche conditions the basis for the establishment of the reservoirs in which *Brucella* persists inside its host. What is currently known of the suitable niches for *Brucella* infection in its various hosts and of the reservoirs of *Brucella* that support chronic infection?

2. Niche and Reservoir

2.1. *Definitions*

The concepts of niche and reservoir may be extrapolated to the infection context. In general, Niche is referred to as the "interrelationship of a species and its relational position in a particular ecosystem including the relationship of the species with the components of the ecosystem itself" [48]. As such, the niche may be influenced by all the factors included in its ecosystem and the niche of a species in a particular ecosystem helps setting up the features of its environment, as the latter ones are crucial for its survival [48]. A specific niche is thus defined as: (i) the specific area where an organism inhabits, (ii) the role or function of an organism or species in an ecosystem, and (iii) the interrelationship of a species with all the factors affecting it. The niche of an organism depicts how it lives and survives as a part of its environment. *Brucella* is a pathogenic genus exceedingly well-adapted to its hosts, which does not survive for extended periods of time in open conditions. This is why it has been called a facultative extracellular intracellular parasite [49]. It also means that *Brucella* has a niche in the intracellular environment of host cells that is specific in a specific cell of the host. This environment sustains extensive replication, in order to facilitate bacterial expansion and subsequent transmission to new host cells, frequently achieved for example, through the heavily infected tissues like the aborted fetus [49].

In the case of Reservoir, the term is referred to an ecologic species that maintain live circulating organisms through the ecosystem over time. For instance, within the niche from a host reservoir, *Brucella* can remain at a low replication rate for a long time, and under favorable conditions, egress to infect other cells and start new replicative cycles. In addition, *Brucella* infections in humans perdure in the ecosystem due to the lack of control of the infection in natural hosts [6,50]. Lymph nodes, spleen, lungs, and the reproductive organs, including placenta, testicular and mammary glands, are well-known target organs for *Brucella* infection [27]. Some predilection for joint articulations has also been reported in human brucellosis [23], meaning that all these organs allow *Brucella* replication and that some may contribute to persistence.

2.2. *Intracellular Niche*

Brucella replicates extensively in the endoplasmic reticulum (ER) compartment within host cells. The mechanisms of entry of the bacterium are still elusive but involve lipid raft-, adhesin- and opsonin-dependent processes [51–54]. After internalization, *Brucella* transits inside the cell engulfed in a phagosome, and multiple virulence factors help the bacteria evade the phagocytic pathway by restricting fusion of the *Brucella* containing vacuole (BCV) with a lysosome. These factors include the cyclic beta-1,2-glucan that operates most probably via cholesterol release [55], the two-component system BvrR/BvrS [56], SepA, which proceeds by excluding the LAMP1 lysosomal protein and preventing the maturation of an active lysosome [57], RicA that regulates vesicle trafficking [58,59] and other possible proteins secreted by the type IV secretion system (T4SS), encoded by the *virB* operon [60,61]. In vitro experiments using macrophage cell lines have indeed shown that the T4SS is required for maturation of the BCV into an ER-like compartment [62]. *Brucella* strains lacking a functional T4SS are unable to escape the fusion with lysosomes, and therefore, highly attenuated in mice and in their natural hosts [63]. Once *Brucella* has impaired phagosome-lysosomal fusion, it replicates in the ER compartment, its replicative niche. Following replication, BCVs interact with host autophagic proteins Beclin1, ULK1,

Atg14, and the IRE1α-UPR signaling axis for bacterial egress and the start of replication cycles within newly infected cells [64–67].

Why does the ER make a good intracellular niche for *Brucella*? The ER is a critical intracellular organelle that not only synthesizes cellular molecules (proteins, lipids, carbohydrates, etc.) but also regulates the transport of the newly synthesized proteins in the exocytic, endocytic and phagosomal pathways. As such, the association of *Brucella* with the host cell ER, like a few successful intracellular pathogens, is expected to be highly beneficial from a nutrient acquisition perspective. Taking advantage of the biosynthetic routes of the host cell, substantial levels of metabolites and nutrients on a local supply base fulfill the complex nutritional requirements of *Brucella* and provide optimal bacterial growth at minimum cost [27,68]. Furthermore, when considering the immune response, localization of *Brucella* in the ER provides an excellent strategy to hide from detection by the immune system and to limit exposure to the cytosolic immune surveillance pathways by avoiding lysosomal fusion for instance. Moreover, the fact that the MHC I peptide loading complex resides in the ER of the two immune cells where *Brucella* replicates, the dendritic cells (DC) and the macrophages, is likely not meaningless. It may suggest a yet unidentified regulatory role of *Brucella* at the level of the setting up of the cross-presentation within the ER and explain its predilection for such cell types.

Indeed, the preference of *Brucella* to replicate within the ER is mostly restricted to phagocytic cells, professional ones as macrophages [69] and DC [70], and non-professional ones, such as placental trophoblasts of pregnant ruminants [35], and fibroblasts [71] or cell lines (Hela) [72]. Even though other cells, including neutrophils [73], lymphocytes [74], and erythrocytes [75], are infectable by *Brucella*, there is no efficient replication inside, and their function is more associated with bacterial dispersion, conferring a regulatory role of these cells in persistence. Of note, if in most cell types *Brucella* replicates within an ER-derived compartment, in extravillous HLA-G+ trophoblasts, *B. abortus*, and *B. suis* fail to reach the "normal" ER-derived niche, in contrast to *B. melitensis*, and replicate within single-membrane acidic lysosomal membrane-associated protein 1 (LAMP1)-positive inclusions [76].

3. Gold Organs in Brucellosis

The "gold organs" for nesting *Brucella*, in which *Brucella* replicates in cells of the reticular endothelial system, include the spleen, lymph nodes, liver, bone marrow, epididymis, and placenta.

3.1. The Reticuloendothelial System

The reticuloendothelial system was originally described in 1924 by K. Aschoff as a group of cells able to incorporate vital dyes from the circulation, "reticulo" referring to their propensity to form a network or reticulum by their cytoplasmic extensions and "endothelial" referring to their vicinity to the endothelium. In 1969, a group of pathologists proposed another term, the monocyte phagocyte system (MPS) [77]. Nowadays the reticuloendothelial system or MPS embraces a family of cells that include committed precursors in the bone marrow, circulating blood monocytes, tissue macrophages, and DC in almost every organ in the body [78].

Brucella has a predilection for organs rich in reticuloendothelial cells (including spleen, liver, bone marrow, and lymph nodes) and is able to replicate successfully in any of them. Intracellular replication is directly linked to *Brucella* pathogenicity and it is not a coincidence that in humans, the most frequent clinical features of brucellosis are an enlarged liver in 65% of the cases, splenomegaly in 52% of the cases (from 40 cases), and lymphadenopathies in children [32,79]. Even in the chicken embryo model, replication of *B. abortus* detected within the rough ER of mesenchymal, mesothelial, and yolk endodermal cells, spreads to all tissues, with the liver and spleen being the most severely infected [80].

In tissues, the typical histopathological response to *Brucella* infection is a granulomatous inflammation, which contains representative members of the MPS, including macrophages with an epithelioid shape, i.e., with an increased amount of cytoplasm. Examination of biopsies from humans and livestock animals reveals granulomas in the liver, spleen, bone marrow, and other tissues [79–82]. As such, the initial replication niche of *Brucella* serves as a platform to establish a chronic infection. *Brucella* infected animals develop granulomatous inflammatory lesions in lymphoid tissues, including the supramammary lymph nodes, reproductive organs, notably the udder, and sometimes joints and synovial membranes. Those granulomas and their intratissular location are responsible for the chronicity of the disease, which can last for months or years [81,83] and in that respect, resemble the granulomas extensively studied in tuberculosis. In fact, in the absence of antibiotic treatment in the acute phase, *Brucella* is able to persist for months without causing significant morbidity or mortality. In the acute phase of infection in a resistant mouse model, the C57BL/6 mice, the formation of granuloma (comprising NOSII+ monocyte-derived inflammatory DC, T cells, and granulocytes) is mediated by MyD88, IL-12, and IFNγ and essential for the control of the bacteria [81,83]. However, these granulomas were not detected in a susceptible murine model of infection, the BALB/c mice, at that stage [81,83]. In *B. melitensis* acutely infected livers, discrete pyogranulomatous inflammatory areas, characterized by a similar influx of neutrophils, macrophages, and monocyte-derived DC, were detected amongst normal hepatocytes in both mouse models [81,83]. At the chronic phase, infected livers displayed well established demarcated infiltration areas of macrophages, lymphocytes, and neutrophils [81]. In chronic granulomas, the presence of lymphocytes is thought to reflect the former activation of the immune system, whereas recruitment of neutrophils suggests that live *Brucella* is still present. The fact that the granuloma areas were typically found surrounding or associated with liver portal tracts and that neutrophils may function as vehicles for dispersion, according to the Trojan horse model [84], supports a dynamic role of granulomas in the development of *Brucella* chronicity. Remarkably, granulomas provide a rich nutrient source, as shown for the dormant non-replicative *Mycobacterium bacilli* that internalize inside the granuloma, lipids from foamy macrophage lipid droplets [85].

3.2. Genital-Reproductive Organs: Placenta and Epididymis

Brucella has a pronounced tropism for genital organs in its natural hosts, placenta in females, and epididymis in males. The placenta is one of the paradisiac organs in terms of replication, containing up to 10^{14} Brucellae in the cow [86,87]. This particular environment allows high replication rates, leading consequently to abortion, the most common clinical feature of brucellosis in livestock. As the main route of infection in these farm animals is aborted fetuses, this seems to be a very efficient strategy to spread *Brucella* progeny to new hosts.

Some common properties in these reproductive organs have shed light on *Brucella's* tropism. Firstly, high concentrations of erythritol are present in uterine, epididymal, and fetal tissues from ruminants [87–90]. Why is this important? Erythritol has been shown to be the preferred carbon/energy source for *Brucella* spp., promoting their massive growth [91]. In addition, the ruminant placenta produces progesterone, which further enhances in vitro *B. abortus* growth [92]. However, *B. abortus* vaccine strain S19 is not stimulated by erythritol [93,94], although it is capable of causing genital infection and abortion [95]. This suggests the existence of other trophic factors. Indeed, the dominance of fructose over glucose takes place in the placenta of cows, sows, ewes, and to a lesser extent in that of other animals [91,96,97]. The same preference applies to the epididymis, seminal fluids, and oviducts of several mammals [91]. As such, both organs play a trophic role and provide effective sources of carbon, nitrogen, and energy for *Brucella* spp. [49,91].

Secondly, the immune-privileged status of the testis and semen, and local immunosuppression at the feto-maternal interface in the placenta might also account for *Brucella* tropism [91].

Thirdly, *Brucella* preferentially replicates within trophoblasts, highly metabolically active cells that adjust their production of proteins and steroids throughout gestation.

Intracellular *Brucella* likely induces the synthesis of steroids and modifies the metabolism of prostaglandin precursors, such as arachidonic acid, which together with the COX-2 enzyme are essential for *Brucella* lymph node persistence and subversion of the immune response [98].

Finally, the high hydrophobicity of the outer-membrane of *Brucella* together with its propensity to replicate within the ER [35,99], may represent an evolutionary adaptation for using hydrophobic substances available within this sub-cellular compartment in trophoblasts [49].

In humans, the genital tropism holds true as *Brucella* induces epididymorchitis [100] and may infect the placenta, even if abortion is very uncommon [76,101].

Therefore, both the localization and abundant multiplication in the reproductive tract of animals is crucial in the biology of this pathogen.

4. Reservoirs

4.1. Bone Marrow

The presence of *Brucella* organisms in humans has been highlighted in the bone marrow in both the acute and chronic phases. In the mouse, the infection remains sequestered within bone marrow cells for a prolonged period of time (until more than 3 months), without significant changes in the bacterial load [102]. For this reason, the bone marrow has recently been proposed as a reservoir [102].

4.1.1. Bone Marrow in the Mouse Model

Murine models have validated bone marrow as an intermittent colonized organ by *Brucella*. Indeed 3 weeks p.i., *Brucella melitensis* has been detected in multiple sites of the murine axial skeleton by in vivo imaging, and immunohistochemistry confirmed its presence in bones, particularly in the lower spine vertebrae, where it preferentially located in a small subset of IBA-1+ monocytes [103]. Similarly, *B. canis* bacterial loads increased in mouse bone marrow from 9 weeks p.i. onward till 12 weeks, a time which coincides with the persistence and chronicity of the infection [104]. Recently, Gutiérrez et al. observed that *Brucella abortus* burden remained constant in bone marrow for up to 168 days p.i., including the acute, chronic, and chronic declining phase in the murine model [102]. However, histopathological alterations varied accordingly to the stage of infection. A granulomatous inflammation, accompanied by augmented numbers of myeloid granulocyte-monocytes progenitors (GMP), granulocytes, and CD4+ lymphocytes, was more severe and diffuse during the acute phase than that of the multifocal chronic phase [102].

The vast number of granulomas in the bone marrow and most importantly their permanence indicates the difficulty for immune cells to eliminate *B. abortus* in such an environment. Interestingly, the cells harboring *Brucella* in higher proportion were granulocytes, monocytes, and GMP progenitors. The fact that monocytes are constantly infected and controlled by *Brucella* strongly suggests that bone marrow is a proper reservoir for persistence in brucellosis.

Moreover, the possibility that *Brucella* infects progenitor stem cells, which lack developed phagocytic machinery, is striking. Recently, it has been shown that hematopoietic stem cells (HSC) sense pathogens, eliciting enhanced myeloid commitment to promoting pathogen clearance of *E. coli* and *Salmonella* Typhimurium [105,106]. In the case of *Brucella*, an increased transient myeloid commitment is triggered by the interaction of *Brucella* Omp25 and host SLAMF1 in the bone marrow, favoring bacterium survival [107]. This may be one of the multiple strategies developed by *Brucella* to promote chronic infection and drive bacterial dissemination.

4.1.2. Bone Marrow in Humans

Bone marrow infection by *Brucella* also occurs in humans in both acute and chronic phases. Usually, a definitive diagnosis of brucellosis is made by culturing *Brucella* from body fluids or tissues [108,109]. The bone marrow has been a recommended tissue to investigate in suspicious cases of brucellosis when the blood culture test is negative. Gotuzzo et al. [110] determined that out of fifty patients, 92% of *Brucella* yielded in bone marrow versus 70% in blood. As mentioned above, granulomas are formed as a reaction to particulate or indigestible agents that persist in tissues for long periods. Additionally, 25% of *Brucella*-infected bone marrows show granulomas together with other pathological changes such as hypercellularity (73% of all cases) or hemophagocytosis (31%) [108,111]. These conglomerates of macrophages, which try to destroy the microbial agent and interact at the same time with lymphocytes might explain rare cases of pancytopenia that are observed in some patients with brucellosis [112].

Human to human brucellosis transmission is extremely rare. In fact, the transmission is due to external factors that promote the transfer of *Brucella*. Bone marrow transplantation facilitates such a transfer, as a concentration of bone marrow cells that are sheltering *Brucella* are transferred to a new host. This means that even in the case of an asymptomatic patient where *Brucella* is present, hidden within bone marrow cells, it will then replicate in the acute phase, or remain there establishing a chronic infection in the recipient host [113,114].

Bones appear to be support structures metabolically inert and resistant to infection by pathogens. However *Brucella* has a tropism for such location and osteoarticular brucellosis is the most common complication in *Brucella*-infected humans (40% from total complications) involving sacroiliitis, spondylitis, and peripheral arthritis [38]. Importantly, osteoarticular lesions are also reported in natural hosts, as infected cattle may exhibit bursitis, arthritis, and hygromas [115,116]. Previous studies have demonstrated the ability of *B. abortus* to invade and replicate within osteoblasts (bone-forming cells), osteocytes (bone-resorbing cells), osteoclasts (multinucleated giant formed by monocyte fusion) [117,118], and in the ER of primary human synoviocytes [119]. Remarkably, osteoclasts originate from the same myeloid precursor cells that give rise to macrophages and myeloid DC and many of the soluble factors (cytokines and growth factors) of immune cells that are regulated by *Brucella*, may regulate the activities of osteoblasts and osteoclasts. A common feature of patients with osteoarticular brucellosis is the presence of leukocyte infiltrates (including monocytes and neutrophils) in the synovial fluid of the joints [120]. The fact that *Brucella* inhibits apoptosis in some of those cells or uses them as a vector to spread to other organs, suggests that its osteoarticular location acts as a reservoir of bacteria to progress towards chronicity.

4.1.3. Bone Marrow Environment

Blood cell formation or hematopoiesis is constantly occurring in the bone marrow, starting from self-replicating-pluripotent HSC, which give rise to multipotent progenitors and further on to lineage-committed cells, such as common lymphoid progenitors (CLP) or common myeloid progenitors (CMP). This sequence of events is tightly regulated by key cytokines and growth factors. The need for such a regulated process suggests the existence of a proper nursing environment called "HSC niche" provided by non-hematopoietic cells and by the undefined architecture of the bone marrow if compared to the spleen or lymph node. Cells forming the niche of HSC have to stay close to the progenitors, produce growth and maintenance of soluble factors, sense signals, and respond to the surrounding environment. This group is made up of endothelial, mesenchymal stem cells, macrophages, DC, granulocytes, megakaryocytes, and lymphocytes. In the HSC niche, amongst others, neutrophils modulate endothelium to produce vascular and HSC regeneration, while DCs control vascularity permeabilization processes [121]. Macrophages physically retain the HSC in the bone marrow and regulate HSC fate by fine-tuning the expression of pro-hematopoietic factors in other cells [122]. Moreover, macrophages via activation of cholesterol sensing LXR receptors eat apoptotic neutrophils that return to the bone marrow to be cleared and allow hematopoietic progenitor release into the blood circulation [123]. Memory CD8+ T lymphocytes and CD4+ Tregs are also localized in the bone marrow and protect HSC and downstream progenitors from various types of stress including inflammation [124].

The bone marrow environment is not only beneficial for HSC, pathogens have also taken advantage of those favorable conditions. *Mycobaterium tuberculosis* is found in mesenchymal cells and HSC of the bone marrow [125,126], which can also host *Listeria monocytogenes*. However, *Listeria* persists rarely in the bone marrow of humans and mouse bones, from where it ultimately seeds the central nervous system [127]. As mentioned before, *Brucella* is able to infect GMP, macrophages, and granulocytes. *Brucella* bone marrow infection is a successful strategy for the following reasons.

(i) Colonization of an immune-privileged organ that lacks intracellular antibacterial mechanisms in early-stage stem cells is easier.
(ii) Expression of drug efflux pumps by bone marrow cells facilitates resistance to antibiotic therapy.
(iii) Cells from bone marrow are highly motile and reside in close proximity to arterial vessels thus offering a wide distribution system for dissemination of the pathogen to target organs.
(iv) *Brucella*'s control of HSC release might explain the splenomegaly observed in human cases as a result of the signal triggered in the bone marrow niche by the apoptotic infected neutrophils to the macrophages, which promotes ensuing cycles of HSC release and extramedullary hematopoiesis.
(v) All *Brucella* preferential target cells differentiate from progenitor stem cells in the bone marrow.

Moreover, the fact that *Brucella* is transmittable by bone marrow transplantation in humans and adoptive transfer experiments in the mouse [107,128], indicates that *Brucella* is able to infect and persist for prolonged periods of time in bone marrow progenitors. It also means that the differentiated infected cells might derive from a GMP, which was already infected and recolonized target tissues in relapse or chronic cases.

4.2. Lymph Nodes

Lymph nodes (LNs) are essential compartments of the immune system, composed of highly organized dynamic leukocyte aggregates, where pathogen defense and adaptive immunity take place [129]. A tightly controlled balance of responses upon challenge results in the induction of either tolerance or immunity [130]. LNs filter fluids from the lymphatic vessels that form an extensive network connecting one LN to another. As such, this network provides conduits for the trafficking of DCs, neutrophils, macrophages, and T cells along the process of immune activation after capturing, transportation, and presentation of antigens in regional lymph nodes [131].

Some bacterial pathogens have exploited the lymphatic system for host colonization in LNs and systemic dissemination [132]. Lymphadenopathy is one of the most common signs of brucellosis in humans, present in approximately more than one-third of cases [4,32,133]. As an example, out of 307 children diagnosed with brucellosis, 112 (36%) exhibited lymphadenopathy [32]. LNs most commonly affected are cervical and axillary LNs due to their proximity to the oral route of infection, the natural route of infection for *Brucella* in humans.

In ruminants, *B. abortus* and *B. melitensis* have a marked affinity for mammary glands and reproductive organs together with the supramammary and genital LNs [134–136]. In goats, about two-thirds of infections acquired naturally during pregnancy lead to infection of the udder and excretion of the bacteria in the milk during successive lactations [137]. In contrast, in sheep, excretion of bacteria in the milk does not last more than two months generally, continuing for up to 140 or 180 days exceptionally [137].

In camels, *B. abortus* and *B. melitensis* have been isolated from LNs, and other organs or compounds like milk, aborted fetus, and placenta [138,139].

The constant or intermittent shedding of *Brucella* in the milk and genital secretions is held by the colonization of LNs and the subsequent spreading through lymphatics. Hence, *Brucellae* have been detected in macrophages [32] and neutrophils in lymphatic draining sites of inoculation or in mammary gland LNs [136]. Since LN leukocytes eventually enter the blood, transportation of infected phagocytes in the lymphatic network might disseminate bacterial organisms throughout the host. In fact, the spleen and the iliac, mammary, and prefemoral LNs are the most reliable samples for isolation purposes in necropsied animals thus proving efficient dissemination [140]. The persistent infection of LNs leads to a constant or intermittent shedding of *Brucella* in the system providing a source of persistent infection in the host and for other animals or humans. Indeed, close contacts between farmers and pastoralists with domesticated animals and the frequent consumption of fresh unpasteurized dairy products increase the maintenance of *Brucella* and the risk of brucellosis in rural and pastoral areas.

4.3. Adipose Tissue

Microorganisms show incredible diversity with respect to which environment they preferentially colonize or invade. Recently, the emergence of new investigation techniques has allowed the analysis of the tropism of several microbes in yet unexamined organs. This is the case for adipose tissue. Adipose tissue is no longer studied as inert lipid storage but as a central regulator of energy homeostasis and immunity. In the past years, fat-associated lymphoid clusters (FALCs) have received much attention [141]. These structures are quite peculiar because contrary to secondary lymphoid organs, FALCs lack both a surrounding capsule and structured compartmentalization of cells. In contrast, they are in direct contact with adipocytes at mucosal surfaces, including omental, mesenteric, mediastinal, gonadal, and pericardial fat [141]. Adipose tissue comprises a vast range of cellular and non-cellular components that support a large network of cells. These include fibroblasts, preadipocytes, cells with mesenchymal and hematopoietic stem cell capacity together with myeloid cells (macrophages, neutrophils, etc.), lymphocytes (T cells, B cells, innate lymphoid cells (ILCs)), eosinophils, mast cells, and NK cells [142]. These immune cells mostly aggregate in clusters or are scattered around them.

The presence of myeloid cell precursors and mature macrophages in the FALC milieu suggests that the lymphoid clusters form permissive microenvironments, where progenitor cells may proliferate locally to generate free macrophages within the cavity where they are located. Another feature of FALCs is that they are rich in vascularization; they are always closely associated with both blood and lymphatic vessels. In terms of response to pathogens, a rapid formation of acute or chronic inflammation occurs in FALCs, suggesting a prominent role of their clusters in the formation of local immune responses.

Many microorganisms, from viruses like HIV and SIV, to bacteria, like *Mycobacterium tuberculosis*, *Ricketssia prowazekii*, *Coxiella burnetti*, and parasites, such as *Plasmodium berghei*, *Trypanozoma cruzi*, and *Trypanozoma brucei*, have been shown to infect adipocytes from humans or mice [143–148]. Their pattern of infection varies according to each pathogen. Some of them establish an intracellular infection specifically within the adipocyte, whereas others remain close to or surround adipocytes and vasculature, or alternatively infect non-fat cells like macrophages.

Not so long ago, nothing was known about *Brucella* and adipocytes or fat tissues. One report in 2018 described the presence of *B. canis* in fat cells of the gastro-splenic ligament, next to lipid droplets and precisely where ER is located, in naturally infected fetuses and neonates [149]. More recently, *B. abortus* has been shown to replicate in a murine fibroblastic derived adipocyte-like cell line, the 3T3-L1 cells, and in its differentiated adipocyte derivative, albeit with less efficiency [150]. However, in this system, bacterial loads start to decline steadily from 3 days p.i. onwards, suggesting that this cell type may not serve as a long-term cellular niche for *Brucella* per se.

Based on these new findings, we would like to propose the adipose tissue in its whole as another singular reservoir for *Brucella* and speculate on the benefits this furtive bacterium might gain from such a location.

(i) Fat tissues are enriched in immune cells as aforementioned. With respect to macrophage populations, M2 polarized ones together with other immune cells recruited locally, like neutrophils, monocytes, and DC, might help to maintain *Brucella* survival and take control of the immune response during infection as suggested by the anti-inflammatory polarization of adipose tissue macrophages potentiated by chronic *T. cruzi* infection [145].

(ii) The majority of drugs are lipophilic compounds, meaning that efficient distribution within the adipose tissue is hampered, as illustrated by some HIV treatment failures. In human brucellosis, 4% of cases undergo to chronic phase [151]. This incidence may rise because of delayed administration of antibiotics or its inefficient delivery to specific organs, such as fat tissue, in the acute phase of infection.

(iii) The localization of fat depots in the perigonadal region of rodents, humans, and ruminants [152] and the preference of *Brucella* to infect epididymis might not just be a simple coincidence and reflect an essential role of adipose tissue in the persistence of this bacteria in reproductive organs. It also suggests that fat tissue provides *Brucella* with a proper environment, i.e., a rich source of nutrients.

(iv) FALCs respond dynamically to stimuli. They expand in size and numbers in response to acute or chronic peritoneal insult [153]. More specifically, an increase of B1 cells, macrophages, and neutrophils recruited via the high endothelial venules has been described in omental FALCs upon injection of *E. coli* LPS or infection [154]. FALCs also support a unique population of CD4+ regulatory cells producers of IL-10 called visceral adipocyte tissue-associated (VAT) Tregs, mostly studied in large fat depots like epididymal fat [155] but also present in the omentum, where it likely regulates local immune responses [156]. Given the essential role of IL-10 in promoting *B. abortus* persistence and pathology [157] and the balance between activation and regulation of immune state that exists in adipocyte tissue, it is conceivable that pathogens, and *Brucella,* in particular, find a golden reservoir here, in which regulatory cells would control the local immune response and create a tolerogenic environment for progression to chronicity.

5. Reservoirs in Wildlife

In addition to the anatomic reservoirs described above, it is imperative to also consider wildlife host reservoirs. The control of brucellosis in humans depends on the control of disease in livestock. However, the creation of new bridges between livestock and wildlife due to human activity is one of the most important factors in disease transmission. As an example, brucellosis has been eradicated in domestic ruminants from most European countries and wild ruminants were not reckoned important hitherto reservoirs. This view changed recently after the notification of two humans cases and the re-emergence of *B. melitensis* in a dairy cattle farm [158], suggesting a possible implication of wildlife as *B. melitensis* infection was identified in a French population of Alpine ibex (*Capra ibex*) [159]. Interestingly, among the 88 seropositive Alpine ibex tested, 58% showed at least one isolation of *B. melitensis* from a urogenital organ (testes, genital tract, urine, or bladder) or a lymph node from the pelvic area (supramammary, internal iliac, and inguinal LNs), meaning that active infection in the pelvic area is at risk of shedding *Brucella* [159]. Moreover, microbiome analysis of bats' guano in India found *B. melitensis* affiliated sequences [160], and *Brucella* sequences were detected in the spleen of two different species of bats from Georgia that were coinfected with *Bartonella* and *Leptospira* [161]. These data indicate that bats may serve as a wildlife reservoir of *Brucella* for grazing goats and sheep.

Regarding other classical *Brucella* species, infections that are recognized as sustainable in wildlife are *B. abortus* in buffalo (*Syncerus caffer*) [162] and bison (*Bison bison*) [163], *B. suis biovar* 2 in wild boar (*Sus Scrofa*), and European hare (*Lepus europeaus*) [164], *B. suis* biovar 4 in reindeer (*Rangifer tarandus*) [165], *Brucella ceti* in cetaceans (*Cetacea*) [166], *Brucella microti* in voles (*Microtus arvalis*) [167] and red fox (*Vulpes vulpes*) [168].

Other species genetically related to an atypical group within the genus have been described in different host species: *Brucella microti*-like in marsh frogs (*Pelophylax ridibundus*) [11]; *Brucella vulpis* in the red fox (*V. vulpes*) [13]; *Brucella inopinata* in White's and Denny's tree frogs, (*Ranoidea caerulea* and *Zhangixalus dennysi*) [169] and humans; *Brucella papionis* in baboons (*Papio* spp.) [15], without mentioning *Brucella* strains isolates from lungworms in porpoise (*Phocoena phocoena*) [170], blue-spotted ribbontail ray (*Taeniura lymma*) [171] and reptile panther chameleon (*Furcifer pardalis*) [172].

Interestingly, *B. microti* positivity was evidenced in mandibular lymph nodes from apparently healthy foxes (*V. vulpes*) assumed to have been contaminated by rodent predation [168].

In 2017, a *B. microti*-like strain was identified in internal organs (heart, lung, spleen, kidney, liver, and reproductive organs) and sometimes in hind limb muscles of marsh frogs in a French farm producing frogs for human consumption [11]. The human pathogenicity of amphibian strains has not been formally demonstrated but cannot be ruled out because the pathogenicity of *B. microti* in wild rodents has been confirmed experimentally in a mouse infection model, with high replication rates in murine macrophages [173].

In cetaceans, *B. ceti* has also been recovered from mandibular, pulmonary, mesenteric, and gastric lymph nodes, spleen, liver, joints, urinary system, and other organs. Noteworthily, Brucellae have been isolated from the female reproductive system, mammary glands, milk and placenta, and in multiple fetal organs, resembling the pathology of terrestrial animals [3,174]. However, it is not the case for pinnipeds (*Pinnipedia*), where most of the isolates came from the spleen, liver, lungs from healthy animals with non-associated pathology. The risk of contamination due to direct contact between coast animals or due to occupational exposure or direct contact with infected aquatic mammals or fomites for humans increases [174].

In both groups, marine mammal parasites, such as lungworms (*Parafilaroides* spp., *Otostrongylus circumlitus*, and *Pseudalius inflexus*), can serve as vectors of marine Brucellae [170,175,176]. In pinnipeds, lungworms are shed in the feces of an infected marine mammal host into the water, then eaten by coprophagic fish; the worms then migrate from the host gastrointestinal tract to their lungs [177], supporting the maintenance and distribution of *Brucella* in aquatic reservoirs.

Recently, *B. pinnipedialis* was recovered for at least 28 days from experimentally infected Atlantic codfish (*Gadus morhua*), suggesting fish as new potential bacterial reservoirs for *Brucella* spp. [178]. Most intriguingly, *B. melitensis* (biovar 3) has been isolated from experimentally and naturally infected Nile catfish (*Clarias gariepinus*) in the delta region of the Nile in Egypt [179]. These findings raise a concern about the role fish and possibly invertebrates may have in transmission or as reservoirs of infection in aquatic environments and for humans, who ingest or handle raw seafood.

B. inopinata was originally described in human infections [12,180]. With the current isolation of *B. inopinata*-like (B13-0095) strain from Pac-Man frogs (*Ceratophyrus ornate*) [181] and the first human case of brucellosis caused by an isolate whose genome is identical, the role of wildlife reservoirs should not be underestimated [182].

Considering that multiple *Brucella* species circulate in totally different hosts and environments in wildlife, control and eradication strategies that had succeeded for livestock contexts require adaptation to wildlife reservoir conditions. For instance, Alpine ibexes and goats after experimental conjunctival vaccination with *B. melitensis* Rev.1 vaccine strain display differences in tissue localization and shedding of the bacteria, as well as humoral immune responses [183]. Likewise, retrospective analysis of the effect of vaccination of elks (*Cervus canadensis*), attending winter feed grounds and adjacent areas of western Wyoming, USA, with the S19 vaccine revealed a failure at reducing post-vaccination seroprevalence of *B. abortus* [184]. Vaccination in wildlife reservoirs involves additional challenges to face, as vaccines need to be validated for safety and efficacy in wild animals. Delivery route and cost being also important issues, its application is by far more complicated than that for livestock [185].

Epidemiological surveillance of brucellosis in terrestrial and marine wildlife reservoirs takes place via several methodologies, amongst which serology is favored [186]. In terrestrial wild animals, if serology is the most commonly applied diagnostic approach, PCR appears to be the most sensitive (36.62% of positive results). Isolation from blood samples and visceral organs constitutes the great majority of specimens used for the detection of *Brucella* spp., noting again lymph nodes as a highly prevalent reservoir (94.6%) [187]. Panorama for surveillance in wildlife is challenging given the diversity of laboratory tests, animal species, environments, cross-reactivity, and non-validation of tests for wildlife [188]. This results in uncertain estimates by serological means of the true prevalence for brucellosis in wildlife, requiring cautious interpretations and other technics when available.

Wildlife reservoirs raise major issues in brucellosis. It is clear that carrying or shedding *Brucella* by wild animals bring a potential risk associated with transmission, persistence, and control in this population as well as domestic ones. However, it is still unknown if wildlife hosts are preferential hosts and if wildlife infections represent a critical reservoir of *Brucella* strains for livestock and therefore humans. New results are needed to elucidate the infection cell cycle of *Brucella* in cells of wildlife and better understand the pathological traits of infection related to disease or persistence.

6. Conclusions

Thanks to its stealthy nature, upon infection, *Brucella* after entering inside a cell, reaches its intracellular niche, the ER, to replicate sheltered from detection by the immune system. This process is central to *Brucella* as it gives the bacterium the ability to maintain replicating-surviving cycles for long periods of time, even at low bacterial numbers, in its cellular niches. Eventually, *Brucella* will take advantage of the environment provided by its anatomic reservoirs, where the cellular niches reside, to disseminate to other organs, where high replication rates can occur. Of course, an organ reservoir would not exist without a pre-existing intracellular niche. It is also generally well interconnected with the other organs of the host to facilitate dissemination. Secretions and products of natural hosts of *Brucella*, livestock, and wildlife, contribute to contamination spreading. Figure 1 illustrates the journey of *Brucella* inside its hosts and recapitulates the intracellular replicative niche, cellular niches, organ reservoirs, and various hosts described in this report. In this challenging time where the world has seen the rapid emergence of a new viral zoonosis transmitted most probably from a pangolin, it is essential to better understand how another zoonosis, such as brucellosis, develops in its numerous wildlife hosts including bats, and livestock ones. Unraveling the molecular and cellular bases of *Brucella* host preference and reservoirs should be continued to preclude opportunities for *Brucella* to jump hosts.

Moreover, the persistence of viable furtive bacteria for extended periods of time highlights the ability of *Brucella* to maintain a chronic state, a feature that complicates brucellosis treatment, control, and eradication programs. A deeper understanding of the different organ reservoirs of *Brucella* should help to design new therapies, which would overcome the inability of current treatments to reach this surreptitious bacterium in certain cells and organs, as is the case for the bone marrow of infected patients.

Brucella niche is distributed in different anatomic reservoirs in the host and especially in some organs, such as the adipose tissue, the role of which in brucellosis is still speculative. This opens up new avenues of research that will undoubtedly contribute to a deeper knowledge of brucellosis and more generally of the mechanisms leading to the chronicity of intracellular pathogens.

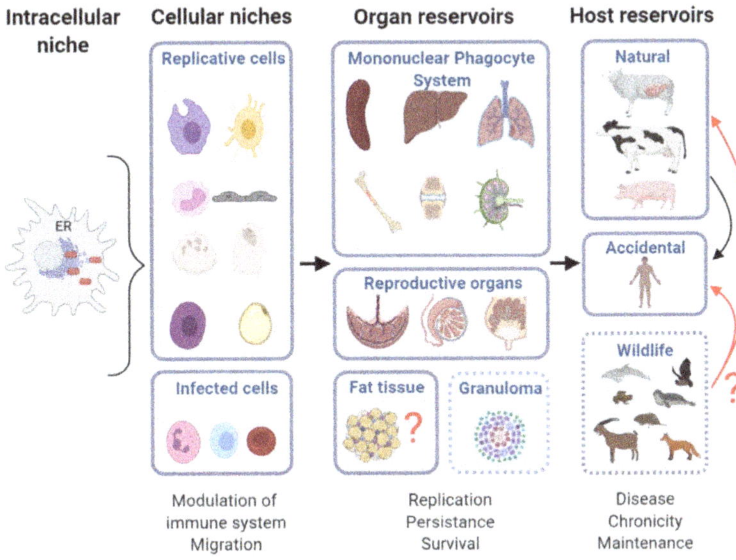

Figure 1. Summary of *Brucella*'s cellular niches and reservoirs. The endoplasmic reticulum is the preferred intracellular niche for *Brucella*, but in some extravillous HLA-G+ trophoblasts, *B. abortus* and *B. suis* are located in lysosomal membrane-associated protein 1 (LAMP1)- and CD63-positive acidic inclusions. *Brucella* replicates in macrophages, dendritic cells, monocytes, trophoblasts, bone cells (osteoclasts, osteoblasts), granulocyte progenitors, adipocytes, and infects other cells such as neutrophils, lymphocytes, and erythrocytes. Some infected cells like neutrophils mediate *Brucella*´s immune response modulation and/or serve as a Trojan horse to disseminate and infect new organs. Several anatomical compartments are populated by *Brucella* infected cells. Organs with high replication rates (placenta, epididymis, mammary glands, lymph nodes, spleen, liver, lungs, and bone marrow) correlate with clinical manifestations of the disease. Once an adaptive immune response is achieved or granulomas contain the infection, *Brucella* develops chronicity and persists at low replication rates. An organ reservoir would not exist without a pre-existing intracellular niche. The structures and physiological characteristics of organ reservoirs allow *Brucella* to start new infection cycles within natural or accidental hosts. Although *Brucella* detection in wild sheep, goats, frogs, fox, bats, and rodents seems almost inconsequential, bacterial loads might be maintained within the host. The zoonotic potential of wildlife reservoirs is still unknown but represents an important risk of transmission to livestock or humans (Created with BioRender.com (accessed on 15 December 2020)).

Author Contributions: Conceptualization, G.G.-E., V.A.-G., S.M. and J.-P.G.; supervision, S.M. and J.-P.G.; writing-original draft preparation, G.G.-E. and S.M.; project administration, S.M. and J.-P.G. All authors have read and agreed to the published version of the manuscript.

Funding: Work is supported by institutional funding from the Centre National de la Recherche Scientifique (CNRS) and the Institut National de la Santé et de la Recherche Médicale (INSERM), and by the Fondation pour la Recherche Médicale (FRM) grant DEQ20170336745, "Laboratories of excellence" programme (Labex INFORM, ANR-11-LABX-0054), the Excellence Initiative of Aix-Marseille Université (AMU) (A*MIDEX), a French "Investissements d'Avenir" programme. GGE is a recipient of o fellowship from the Fondation de Coopération Scientifique "Méditerranée Infection" and from Campus France.

Acknowledgments: Authors thank all members of the JPG lab for their support.

Conflicts of Interest: The authors declare no conflict of interest.

References

1. Fouskis, I.; Sandalakis, V.; Christidou, A.; Tsatsaris, A.; Tzanakis, N.; Tselentis, Y.; Psaroulaki, A. The epidemiology of Brucellosis in Greece, 2007–2012: A 'One Health' approach. *Trans. R. Soc. Trop. Med. Hyg.* **2018**, *112*, 124–135. [CrossRef]
2. Guzmán-Hernández, R.L.; Contreras-Rodríguez, A.; Ávila-Calderón, E.D.; Morales-García, M.R. Brucelosis: Zoonosis de importancia en México. *Rev. Chil. Infectol.* **2016**, *33*, 656–662. [CrossRef] [PubMed]
3. Hernández-Mora, G.; Bonilla-Montoya, R.; Barrantes-Granados, O.; Esquivel-Suárez, A.; Montero-Caballero, D.; González-Barrientos, R.; Fallas-Monge, Z.; Palacios-Alfaro, J.D.; Baldi, M.; Campos, E.; et al. Brucellosis in mammals of Costa Rica: An epidemiological survey. *PLoS ONE* **2017**, *12*, e0182644. [CrossRef] [PubMed]
4. Zheng, R.; Xie, S.; Lu, X.; Sun, L.; Zhou, Y.; Zhang, Y.; Wang, K. A Systematic Review and Meta-Analysis of Epidemiology and Clinical Manifestations of Human Brucellosis in China. *Biomed. Res. Int.* **2018**, *2018*, 1–10. [CrossRef] [PubMed]
5. Franco, M.P.; Mulder, M.; Gilman, R.H.; Smits, H.L. Human brucellosis. *Lancet Infect. Dis.* **2007**, *7*, 775–786. [CrossRef]
6. O'Callaghan, D. Human brucellosis: Recent advances and future challenges. *Infect. Dis. Poverty* **2020**, *9*, 101. [CrossRef]
7. Bardhan, D.; Kumar, S.; Verma, M.R.; Bangar, Y.C. Economic losses due to brucellosis in India. *Indian J. Comp. Microbiol. Immunol. Infect. Dis.* **2020**, *41*, 19. [CrossRef]
8. Zeng, J.Y.; Robertson, I.D.; Ji, Q.M.; Dawa, Y.L.; Bruce, M. Evaluation of the economic impact of brucellosis in domestic yaks of Tibet. *Transbound. Emerg. Dis.* **2019**, *66*, 476–487. [CrossRef]
9. Suárez-Esquivel, M.; Chaves-Olarte, E.; Moreno, E.; Guzmán-Verri, C. *Brucella* Genomics: Macro and Micro Evolution. *Int. J. Mol. Sci.* **2020**, *21*, 7749. [CrossRef]
10. Foster, G.; Osterman, B.S.; Godfroid, J.; Jacques, I.; Cloeckaert, A. *Brucella ceti* sp. nov. and *Brucella pinnipedialis* sp. nov. for *Brucella* strains with cetaceans and seals as their preferred hosts. *Int. J. Syst. Evol. Microbiol.* **2007**, *57*, 2688–2693. [CrossRef]
11. Jaÿ, M.; Freddi, L.; Mick, V.; Durand, B.; Girault, G.; Perrot, L.; Taunay, B.; Vuilmet, T.; Azam, D.; Ponsart, C.; et al. *Brucella microti*-like prevalence in French farms producing frogs. *Transbound. Emerg. Dis.* **2020**, *67*, 617–625. [CrossRef] [PubMed]
12. Scholz, H.C.; Nöckler, K.; Göllner, C.; Bahn, P.; Vergnaud, G.; Tomaso, H.; Al Dahouk, S.; Kämpfer, P.; Cloeckaert, A.; Maquart, M.; et al. *Brucella inopinata* sp. nov., isolated from a breast implant infection. *Int. J. Syst. Evol. Microbiol.* **2010**, *60*, 801–808. [CrossRef] [PubMed]
13. Scholz, H.C.; Revilla-Fernández, S.; Dahouk, S.A.; Hammerl, J.A.; Zygmunt, M.S.; Cloeckaert, A.; Koylass, M.; Whatmore, A.M.; Blom, J.; Vergnaud, G.; et al. *Brucella vulpis* sp. nov., isolated from mandibular lymph nodes of red foxes (*Vulpes vulpes*). *Int. J. Syst. Evol. Microbiol.* **2016**, *66*, 2090–2098. [CrossRef] [PubMed]
14. Suárez-Esquivel, M.; Baker, K.S.; Ruiz-Villalobos, N.; Hernández-Mora, G.; Barquero-Calvo, E.; González-Barrientos, R.; Castillo-Zeledón, A.; Jiménez-Rojas, C.; Chacón-Díaz, C.; Cloeckaert, A.; et al. *Brucella* Genetic Variability in Wildlife Marine Mammals Populations Relates to Host Preference and Ocean Distribution. *Genome Biol. Evol.* **2017**, *9*, 1901–1912. [CrossRef]
15. Whatmore, A.M.; Davison, N.; Cloeckaert, A.; Al Dahouk, S.; Zygmunt, M.S.; Brew, S.D.; Perrett, L.L.; Koylass, M.S.; Vergnaud, G.; Quance, C.; et al. *Brucella papionis* sp. nov., isolated from baboons (*Papio* spp.). *Int. J. Syst. Evol. Microbiol.* **2014**, *64*, 4120–4128. [CrossRef]
16. Whatmore, A.M.; Dale, E.J.; Stubberfield, E.; Muchowski, J.; Koylass, M.; Dawson, C.; Gopaul, K.K.; Perrett, L.L.; Jones, M.; Lawrie, A. Isolation of *Brucella* from a White's tree frog (*Litoria caerulea*). *JMM Case Rep.* **2015**, 2. [CrossRef]
17. Godfroid, J.; Al Dahouk, S.; Pappas, G.; Roth, F.; Matope, G.; Muma, J.; Marcotty, T.; Pfeiffer, D.; Skjerve, E. A "One Health" surveillance and control of brucellosis in developing countries: Moving away from improvisation. *Comp. Immunol. Microbiol. Infect. Dis.* **2013**, *36*, 241–248. [CrossRef]
18. Ay, N.; Kaya, S.; Anil, M.; Alp, V.; Beyazit, U.; Yuksel, E.; Danis, R. Pulmonary Involvement in Brucellosis, a Rare Complication of Renal Transplant: Case Report and Brief Review. *Exp. Clin. Transplant. Off. J. Middle East. Soc. Organ. Transplant.* **2018**, *16*, 757–760. [CrossRef]
19. Bosilkovski, M.; Arapović, J.; Keramat, F. Human brucellosis in pregnancy—An overview. *Bosn. J. Basic Med. Sci.* **2019**. [CrossRef] [PubMed]
20. Ting, I.-W.; Ho, M.-W.; Sung, Y.-J.; Tien, N.; Chi, C.-Y.; Ho, H.-C.; Huang, C.-C. Brucellosis in a renal transplant recipient. *Transpl. Infect. Dis.* **2013**. [CrossRef] [PubMed]
21. Tuon, F.F.; Gondolfo, R.B.; Cerchiari, N. Human-to-human transmission of *Brucella*—A systematic review. *Trop. Med. Int. Health* **2017**, *22*, 539–546. [CrossRef]
22. Wallach, J.C.; Samartino, L.E.; Efron, A.; Baldi, P.C. Human infection by *Brucella melitensis*: An outbreak attributed to contact with infected goats. *FEMS Immunol. Med. Microbiol.* **1997**, *19*, 315–321. [CrossRef]
23. Aygen, B.; Doğanay, M.; Sümerkan, B.; Yildiz, O.; Kayabaş, Ü. Clinical manifestations, complications and treatment of brucellosis: A retrospective evaluation of 480 patients. *Médecine Mal. Infect.* **2002**, *32*, 485–493. [CrossRef]
24. Buzgan, T.; Karahocagil, M.K.; Irmak, H.; Baran, A.I.; Karsen, H.; Evirgen, O.; Akdeniz, H. Clinical manifestations and complications in 1028 cases of brucellosis: A retrospective evaluation and review of the literature. *Int. J. Infect. Dis.* **2010**, *14*, e469–e478. [CrossRef]
25. Ko, J.; Splitter, G.A. Molecular Host-Pathogen Interaction in Brucellosis: Current Understanding and Future Approaches to Vaccine Development for Mice and Humans. *Clin. Microbiol. Rev.* **2003**, *16*, 65–78. [CrossRef]
26. Köse, Ş.; Serin Senger, S.; Akkoçlu, G.; Kuzucu, L.; Ulu, Y.; Ersan, G.; Oğuz, F. Clinical manifestations, complications, and treatment of brucellosis: Evaluation of 72 cases. *Turk. J. Med. Sci.* **2014**, *44*, 220–223. [CrossRef] [PubMed]

27. Moreno, E.; Moriyón, I. The Genus *Brucella*. In *The Prokaryotes*; Dworkin, M., Falkow, S., Rosenberg, E., Schleifer, K.-H., Stackebrandt, E., Eds.; Springer: New York, NY, USA, 2006; pp. 315–456.
28. Mcdermott, J.J.; Grace, D.; Zinsstag, J. Economics of brucellosis impact and control in low-income countries. *Rev. Sci. Tech. De L'oie* **2013**, *32*, 249–261. [CrossRef] [PubMed]
29. Olsen, S.C.; Palmer, M.V. Advancement of Knowledge of *Brucella* over the Past 50 Years. *Vet. Pathol.* **2014**, *51*, 1076–1089. [CrossRef]
30. Moreno, E. Retrospective and prospective perspectives on zoonotic brucellosis. *Front. Microbiol.* **2014**, *5*. [CrossRef]
31. Poester, F.P.; Samartino, L.E.; Santos, R.L. Pathogenesis and pathobiology of brucellosis in livestock. *Rev. Sci. Tech. De L'oie* **2013**, *32*, 105–115. [CrossRef]
32. von Bargen, K.; Gagnaire, A.; Arce-Gorvel, V.; de Bovis, B.; Baudimont, F.; Chasson, L.; Bosilkovski, M.; Papadopoulos, A.; Martirosyan, A.; Henri, S.; et al. Cervical Lymph Nodes as a Selective Niche for *Brucella* during Oral Infections. *PLoS ONE* **2015**, *10*, e0121790. [CrossRef]
33. Moreno, E.; Barquero-Calvo, E. The Role of Neutrophils in Brucellosis. *Microbiol. Mol. Biol. Rev.* **2020**, *84*, e00048-20. [CrossRef]
34. Anderson, T.D.; Meador, V.P.; Cheville, N.F. Pathogenesis of Placentitis in the Goat Inoculated with *Brucella abortus*. I. Gross and Histologic Lesions. *Vet. Pathol.* **1986**, *23*, 219–226. [CrossRef]
35. Meador, V.P.; Deyoe, B.L. Intracellular Localization of *Brucella abortus* in Bovine Placenta. *Vet. Pathol.* **1989**, *26*, 513–515. [CrossRef]
36. Tobias, L.; Cordes, D.O.; Schurig, G.G. Placental Pathology of the Pregnant Mouse Inoculated with *Brucella abortus* Strain 2308. *Vet. Pathol.* **1993**, *30*, 119–129. [CrossRef] [PubMed]
37. Harmon, B.G.; Adams, L.G.; Frey, M. Survival of rough and smooth strains of *Brucella abortus* in bovine mammary gland macrophages. *Am. J. Vet. Res.* **1988**, *49*, 1092–1097. [PubMed]
38. Hull, N.C.; Schumaker, B.A. Comparisons of brucellosis between human and veterinary medicine. *Infect. Ecol. Epidemiol.* **2018**, *8*, 1500846. [CrossRef] [PubMed]
39. Amjadi, O.; Rafiei, A.; Mardani, M.; Zafari, P.; Zarifian, A. A review of the immunopathogenesis of Brucellosis. *Infect. Dis.* **2019**, *51*, 321–333. [CrossRef] [PubMed]
40. Pabuccuoglu, O.; Ecemis, T.; El, S.; Coskun, A.; Akcali, S.; Sanlidag, T. Evaluation of serological tests for diagnosis of brucellosis. *Jpn. J. Infect. Dis.* **2011**, *64*, 272–276. [PubMed]
41. *Brucellosis in Humans and Animals*; Weltgesundheitsorganisation; FAO (Eds.) WHO: Geneva, Switzerland, 2006; p. 89.
42. Megid, J.; Antonio Mathias, L.; Robles, C. Clinical Manifestations of Brucellosis in Domestic Animals and Humans. *Open Vet. Sci. J.* **2010**, *4*, 119–126. [CrossRef]
43. Barquero-Calvo, E.; Chaves-Olarte, E.; Weiss, D.S.; Guzmán-Verri, C.; Chacón-Díaz, C.; Rucavado, A.; Moriyón, I.; Moreno, E. *Brucella abortus* Uses a Stealthy Strategy to Avoid Activation of the Innate Immune System during the Onset of Infection. *PLoS ONE* **2007**, *2*, e631. [CrossRef] [PubMed]
44. Martirosyan, A.; Moreno, E.; Gorvel, J.-P. An evolutionary strategy for a stealthy intracellular *Brucella* pathogen: *Brucella* pathogenenesis. *Immunol. Rev.* **2011**, *240*, 211–234. [CrossRef] [PubMed]
45. Guzman-Verri, C.; Manterola, L.; Sola-Landa, A.; Parra, A.; Cloeckaert, A.; Garin, J.; Gorvel, J.-P.; Moriyon, I.; Moreno, E.; Lopez-Goni, I. The two-component system BvrR/BvrS essential for *Brucella abortus* virulence regulates the expression of outer membrane proteins with counterparts in members of the Rhizobiaceae. *Proc. Natl. Acad. Sci. USA* **2002**, *99*, 12375–12380. [CrossRef] [PubMed]
46. Pizarro-Cerdá, J.; Desjardins, M.; Moreno, E.; Akira, S.; Gorvel, J.P. Modulation of endocytosis in nuclear factor IL-6($-/-$) macrophages is responsible for a high susceptibility to intracellular bacterial infection. *J. Immunol.* **1999**, *162*, 3519–3526. [PubMed]
47. Gomes, M.T.R.; Campos, P.C.; de Almeida, L.A.; Oliveira, F.S.; Costa, M.M.S.; Marim, F.M.; Pereira, G.S.M.; Oliveira, S.C. The role of innate immune signals in immunity to *Brucella abortus*. *Front. Cell. Infect. Microbiol.* **2012**, *2*. [CrossRef]
48. Polechová, J.; Storch, D. Ecological Niche. In *Encyclopedia of Ecology*; Elsevier: Amsterdam, The Netherlands, 2019; pp. 72–80.
49. Gorvel, J.P.; Moreno, E. *Brucella* intracellular life: From invasion to intracellular replication. *Vet. Microbiol.* **2002**, *90*, 281–297. [CrossRef]
50. Godfroid, J.; Garin-Bastuji, B.; Saegerman, C.; Blasco, J.M. Brucellosis in terrestrial wildlife. *Rev. Sci. Tech. De L'oie* **2013**, *32*, 27–42. [CrossRef]
51. Kim, S.; Watarai, M.; Suzuki, H.; Makino, S.-i.; Kodama, T.; Shirahata, T. Lipid raft microdomains mediate class A scavenger receptor-dependent infection of *Brucella abortus*. *Microb. Pathog.* **2004**, *37*, 11–19. [CrossRef]
52. Naroeni, A.; Porte, F. Role of Cholesterol and the Ganglioside GM1 in Entry and Short-Term Survival of *Brucella suis* in Murine Macrophages. *Infect. Immun.* **2002**, *70*, 1640–1644. [CrossRef]
53. Ruiz-Ranwez, V.; Posadas, D.M.; Van der Henst, C.; Estein, S.M.; Arocena, G.M.; Abdian, P.L.; Martín, F.A.; Sieira, R.; De Bolle, X.; Zorreguieta, A. BtaE, an adhesin that belongs to the trimeric autotransporter family, is required for full virulence and defines a specific adhesive pole of *Brucella suis*. *Infect. Immun.* **2013**, *81*, 996–1007. [CrossRef]
54. Ruiz-Ranwez, V.; Posadas, D.M.; Estein, S.M.; Abdian, P.L.; Martin, F.A.; Zorreguieta, A. The BtaF Trimeric Autotransporter of *Brucella suis* Is Involved in Attachment to Various Surfaces, Resistance to Serum and Virulence. *PLoS ONE* **2013**, *8*, e79770. [CrossRef] [PubMed]
55. Arellano-Reynoso, B.; Lapaque, N.; Salcedo, S.; Briones, G.; Ciocchini, A.E.; Ugalde, R.; Moreno, E.; Moriyón, I.; Gorvel, J.-P. Cyclic β-1,2-glucan is a *Brucella* virulence factor required for intracellular survival. *Nat. Immunol.* **2005**, *6*, 618–625. [CrossRef]

56. Sola-Landa, A.; Pizarro-Cerdá, J.; Grilló, M.J.; Moreno, E.; Moriyón, I.; Blasco, J.M.; Gorvel, J.P.; López-Goñi, I. A two-component regulatory system playing a critical role in plant pathogens and endosymbionts is present in *Brucella abortus* and controls cell invasion and virulence. *Mol. Microbiol.* **1998**, *29*, 125–138. [CrossRef] [PubMed]
57. Döhmer, P.H.; Valguarnera, E.; Czibener, C.; Ugalde, J.E. Identification of a type IV secretion substrate of *B rucella abortus* that participates in the early stages of intracellular survival: A novel *Brucella virB* secretion substrate. *Cell. Microbiol.* **2014**, *16*, 396–410. [CrossRef] [PubMed]
58. De Barsy, M.; Jamet, A.; Filopon, D.; Nicolas, C.; Laloux, G.; Rual, J.-F.; Muller, A.; Twizere, J.-C.; Nkengfac, B.; Vandenhaute, J.; et al. Identification of a *Brucella* spp. secreted effector specifically interacting with human small GTPase Rab2: A *Brucella* effector interacting with Rab2. *Cell. Microbiol.* **2011**, *13*, 1044–1058. [CrossRef]
59. Smith, E.P.; Cotto-Rosario, A.; Borghesan, E.; Held, K.; Miller, C.N.; Celli, J. Epistatic Interplay between Type IV Secretion Effectors Engages the Small GTPase Rab2 in the *Brucella* Intracellular Cycle. *mBio* **2020**, *11*, e03350-19. [CrossRef]
60. Celli, J.; de Chastellier, C.; Franchini, D.-M.; Pizarro-Cerda, J.; Moreno, E.; Gorvel, J.-P. *Brucella* Evades Macrophage Killing via VirB-dependent Sustained Interactions with the Endoplasmic Reticulum. *J. Exp. Med.* **2003**, *198*, 545–556. [CrossRef] [PubMed]
61. O'Callaghan, D.; Cazevieille, C.; Allardet-Servent, A.; Boschiroli, M.L.; Bourg, G.; Foulongne, V.; Frutos, P.; Kulakov, Y.; Ramuz, M. A homologue of the *Agrobacterium tumefaciens* VirB and *Bordetella pertussis* Ptl type IV secretion systems is essential for intracellular survival of *Brucella suis*: *Brucella* type IV secretion. *Mol. Microbiol.* **2002**, *33*, 1210–1220. [CrossRef] [PubMed]
62. Comerci, D.J.; Martinez-Lorenzo, M.J.; Sieira, R.; Gorvel, J.-P.; Ugalde, R.A. Essential role of the VirB machinery in the maturation of the *Brucella abortus*-containing vacuole. *Cell. Microbiol.* **2001**, *3*, 159–168. [CrossRef] [PubMed]
63. Hong, P.C.; Tsolis, R.M.; Ficht, T.A. Identification of Genes Required for Chronic Persistence of *Brucella abortus* in Mice. *Infect. Immun.* **2000**, *68*, 4102–4107. [CrossRef]
64. Byndloss, M.X.; Tsai, A.Y.; Walker, G.T.; Miller, C.N.; Young, B.M.; English, B.C.; Seyffert, N.; Kerrinnes, T.; de Jong, M.F.; Atluri, V.L.; et al. *Brucella abortus* Infection of Placental Trophoblasts Triggers Endoplasmic Reticulum Stress-Mediated Cell Death and Fetal Loss via Type IV Secretion System-Dependent Activation of CHOP. *mBio* **2019**, *10*, e01538-19. [CrossRef]
65. Celli, J. The changing nature of the *B rucella*-containing vacuole: The *Brucella* -containing vacuole. *Cell. Microbiol.* **2015**, *17*, 951–958. [CrossRef] [PubMed]
66. Pandey, A.; Lin, F.; Cabello, A.L.; da Costa, L.F.; Feng, X.; Feng, H.-Q.; Zhang, M.-Z.; Iwawaki, T.; Rice-Ficht, A.; Ficht, T.A.; et al. Activation of Host IRE1α-Dependent Signaling Axis Contributes the Intracellular Parasitism of *Brucella melitensis*. *Front. Cell. Infect. Microbiol.* **2018**, *8*, 103. [CrossRef] [PubMed]
67. Starr, T.; Child, R.; Wehrly Tara, D.; Hansen, B.; Hwang, S.; López-Otin, C.; Virgin Herbert, W.; Celli, J. Selective Subversion of Autophagy Complexes Facilitates Completion of the *Brucella* Intracellular Cycle. *Cell Host Microbe* **2012**, *11*, 33–45. [CrossRef]
68. Roy, C.R.; Salcedo, S.P.; Gorvel, J.-P.E. Pathogen–endoplasmic-reticulum interactions: In through the out door. *Nat. Rev. Immunol.* **2006**, *6*, 136–147. [CrossRef]
69. Celli, J. Surviving inside a macrophage: The many ways of *Brucella*. *Res. Microbiol.* **2006**, *157*, 93–98. [CrossRef] [PubMed]
70. Salcedo, S.P.; Marchesini, M.I.; Lelouard, H.; Fugier, E.; Jolly, G.; Balor, S.; Muller, A.; Lapaque, N.; Demaria, O.; Alexopoulou, L.; et al. *Brucella* Control of Dendritic Cell Maturation Is Dependent on the TIR-Containing Protein Btp1. *PLoS Pathog.* **2008**, *4*, e21. [CrossRef] [PubMed]
71. Demars, A.; Lison, A.; Machelart, A.; Van Vyve, M.; Potemberg, G.; Vanderwinden, J.-M.; De Bolle, X.; Letesson, J.-J.; Muraille, E. Route of Infection Strongly Impacts the Host-Pathogen Relationship. *Front. Immunol.* **2019**, *10*, 1589. [CrossRef] [PubMed]
72. Pizarro-Cerdá, J.; Méresse, S.; Parton, R.G.; van der Goot, G.; Sola-Landa, A.; Lopez-Goñi, I.; Moreno, E.; Gorvel, J.-P. *Brucella abortus* Transits through the Autophagic Pathway and Replicates in the Endoplasmic Reticulum of Nonprofessional Phagocytes. *Infect. Immun.* **1998**, *66*, 5711–5724. [CrossRef] [PubMed]
73. Barquero-Calvo, E.; Mora-Cartín, R.; Arce-Gorvel, V.; de Diego, J.L.; Chacón-Díaz, C.; Chaves-Olarte, E.; Guzmán-Verri, C.; Buret, A.G.; Gorvel, J.-P.; Moreno, E. *Brucella abortus* Induces the Premature Death of Human Neutrophils through the Action of Its Lipopolysaccharide. *PLoS Pathog.* **2015**, *11*, e1004853. [CrossRef]
74. Goenka, R.; Guirnalda, P.D.; Black, S.J.; Baldwin, C.L. B Lymphocytes Provide an Infection Niche for Intracellular Bacterium *Brucella abortus*. *J. Infect. Dis.* **2012**, *206*, 91–98. [CrossRef]
75. Vitry, M.-A.; Hanot Mambres, D.; Deghelt, M.; Hack, K.; Machelart, A.; Lhomme, F.; Vanderwinden, J.-M.; Vermeersch, M.; De Trez, C.; Pérez-Morga, D.; et al. *Brucella melitensis* invades murine erythrocytes during infection. *Infect. Immun.* **2014**, *82*, 3927–3938. [CrossRef] [PubMed]
76. Salcedo, S.P.; Chevrier, N.; Lacerda, T.L.S.; Ben Amara, A.; Gerart, S.; Gorvel, V.A.; de Chastellier, C.; Blasco, J.M.; Mege, J.-L.; Gorvel, J.-P. Pathogenic Brucellae Replicate in Human Trophoblasts. *J. Infect. Dis.* **2013**, *207*, 1075–1083. [CrossRef]
77. Yona, S.; Gordon, S. From the Reticuloendothelial to Mononuclear Phagocyte System—The Unaccounted Years. *Front. Immunol.* **2015**, *6*. [CrossRef]
78. Summers, K.M.; Bush, S.J.; Hume, D.A. Network analysis of transcriptomic diversity amongst resident tissue macrophages and dendritic cells in the mouse mononuclear phagocyte system. *PLoS Biol.* **2020**, *18*, e3000859. [CrossRef]
79. Cervantes, F.; Carbonell, J.; Bruguera, M.; Force, L.; Webb, S. Liver disease in brucellosis. A clinical and pathological study of 40 cases. *Postgrad. Med. J.* **1982**, *58*, 346–350. [CrossRef]
80. Detilleux, P.G.; Cheville, N.F.; Deyoe, B.L. Pathogenesis of *Brucella abortus* in Chicken Embryos. *Vet. Pathol.* **1988**, *25*, 138–146. [CrossRef] [PubMed]

81. Daggett, J.; Rogers, A.; Harms, J.; Splitter, G.A.; Durward-Diioia, M. Hepatic and splenic immune response during acute vs. chronic *Brucella melitensis* infection using in situ microscopy. *Comp. Immunol. Microbiol. Infect. Dis.* **2020**, *73*, 101490. [CrossRef]
82. Karim, M.F.; Maruf, A.A.; Yeasmin, F.; Shafy, N.M.; Khan, A.H.N.A.; Rahman, A.K.M.A.; Bhuiyan, M.J.S.; Hasan, M.M.; Karim, M.R.; Hasan, M.T.; et al. Histopathological changes of brucellosis in experimentally infected guinea pig. *Bangladesh J. Vet. Med.* **2019**, *17*. [CrossRef]
83. Copin, R.; Vitry, M.-A.; Hanot Mambres, D.; Machelart, A.; De Trez, C.; Vanderwinden, J.-M.; Magez, S.; Akira, S.; Ryffel, B.; Carlier, Y.; et al. In Situ Microscopy Analysis Reveals Local Innate Immune Response Developed around *Brucella* Infected Cells in Resistant and Susceptible Mice. *PLoS Pathog.* **2012**, *8*, e1002575. [CrossRef] [PubMed]
84. Gutiérrez-Jiménez, C.; Mora-Cartín, R.; Altamirano-Silva, P.; Chacón-Díaz, C.; Chaves-Olarte, E.; Moreno, E.; Barquero-Calvo, E. Neutrophils as Trojan Horse Vehicles for *Brucella abortus* Macrophage Infection. *Front. Immunol.* **2019**, *10*, 1012. [CrossRef]
85. Peyron, P.; Vaubourgeix, J.; Poquet, Y.; Levillain, F.; Botanch, C.; Bardou, F.; Daffé, M.; Emile, J.-F.; Marchou, B.; Cardona, P.-J.; et al. Foamy Macrophages from Tuberculous Patients' Granulomas Constitute a Nutrient-Rich Reservoir for *M. tuberculosis* Persistence. *Plos Pathog.* **2008**, *4*, e1000204. [CrossRef] [PubMed]
86. Alexander, B.; Schnurrenberger, P.R.; Brown, R.R. Numbers of *Brucella abortus* in the placenta, umbilicus and fetal fluid of two naturally infected cows. *Vet. Rec.* **1981**, *108*, 500. [CrossRef]
87. Smith, H.; Williams, A.E.; Pearce, J.H.; Keppie, J.; Harris-Smith, P.W.; Fitz-George, R.B.; Witt, K. Fœtal Erythritol: A Cause of the Localization of *Brucella abortus* in Bovine Contagious Abortion. *Nature* **1962**, *193*, 47–49. [CrossRef] [PubMed]
88. Clark, J.B.K.; Graham, E.F.; Lewis, B.A.; Smith, F. D-mannitol, erythritol and glycerol in bovine semen. *Reproduction* **1967**, *13*, 189–197. [CrossRef]
89. Essenberg, R.C.; Seshadri, R.; Nelson, K.; Paulsen, I. Sugar metabolism by Brucellae. *Vet. Microbiol.* **2002**, *90*, 249–261. [CrossRef]
90. Keppie, J.; Williams, A.E.; Witt, K.; Smith, H. The role of erythritol in the tissue localization of the brucellae. *Br. J. Exp. Pathol.* **1965**, *46*, 104–108.
91. Letesson, J.-J.; Barbier, T.; Zúñiga-Ripa, A.; Godfroid, J.; De Bolle, X.; Moriyón, I. *Brucella* Genital Tropism: What's on the Menu. *Front. Microbiol.* **2017**, *8*. [CrossRef]
92. Samartino, L.E.; Enright, F.M. Pathogenesis of abortion of bovine brucellosis. *Comp. Immunol. Microbiol. Infect. Dis.* **1993**, *16*, 95–101. [CrossRef]
93. Keppie, J.; Witt, K.; Smith, H. The Effect of Erythritol on the Growth of S19 and other Attenuated Strains of *Brucella abortus*. *Res. Vet. Sci.* **1967**, *8*, 294–296. [CrossRef]
94. Rodríguez, M.C.; Viadas, C.; Seoane, A.; Sangari, F.J.; López-Goñi, I.; García-Lobo, J.M. Evaluation of the Effects of Erythritol on Gene Expression in *Brucella abortus*. *PLoS ONE* **2012**, *7*, e50876. [CrossRef]
95. Corner, L.A.; Alton, G.G. Persistence of *Brucella abortus* strain 19 infection in adult cattle vaccinated with reduced doses. *Res. Vet. Sci.* **1981**, *31*, 342–344. [CrossRef]
96. Alexander, D.P.; Huggett, A.S.G.; Nixon, D.A.; Widdas, W.F. The placental transfer of sugars in the sheep: The influence of concentration gradient upon the rates of hexose formation as shown in umbilical perfusion of the placenta. *J. Physiol.* **1955**, *129*, 367–383. [CrossRef]
97. Hastein, T.; Velle, W. Placental aldose reductase activity and foetal blood fructose during bovine pregnancy. *Reproduction* **1968**, *15*, 47–52. [CrossRef]
98. Gagnaire, A.; Gorvel, L.; Papadopoulos, A.; Von Bargen, K.; Mège, J.-L.; Gorvel, J.-P. COX-2 Inhibition Reduces *Brucella* Bacterial Burden in Draining Lymph Nodes. *Front. Microbiol.* **2016**, *7*, 1987. [CrossRef]
99. Anderson, T.D.; Cheville, N.F.; Meador, V.P. Pathogenesis of Placentitis in the Goat Inoculated with *Brucella abortus*. II. Ultrastructural Studies. *Vet. Pathol.* **1986**, *23*, 227–239. [CrossRef]
100. Queipo-Ortuño, M.I.; Colmenero, J.D.; Muñoz, N.; Baeza, G.; Clavijo, E.; Morata, P. Rapid Diagnosis of *Brucella* Epididymo-Orchitis by Real-Time Polymerase Chain Reaction Assay in Urine Samples. *J. Urol.* **2006**, *176*, 2290–2293. [CrossRef] [PubMed]
101. Al-Tawfiq, J.; Memish, Z. Pregnancy Associated Brucellosis. *Recent Pat. Anti-Infect. Drug Discov.* **2013**, *8*, 47–50. [CrossRef] [PubMed]
102. Gutiérrez-Jiménez, C.; Hysenaj, L.; Alfaro-Alarcón, A.; Mora-Cartín, R.; Arce-Gorvel, V.; Moreno, E.; Gorvel, J.P.; Barquero-Calvo, E. Persistence of *Brucella abortus* in the Bone Marrow of Infected Mice. *J. Immunol. Res.* **2018**, *2018*, 1–8. [CrossRef]
103. Magnani, D.M.; Lyons, E.T.; Forde, T.S.; Shekhani, M.T.; Adarichev, V.A.; Splitter, G.A. Osteoarticular tissue infection and development of skeletal pathology in murine brucellosis. *Dis. Models Mech.* **2013**, *6*, 811–818. [CrossRef] [PubMed]
104. Chacón-Díaz, C.; Altamirano-Silva, P.; González-Espinoza, G.; Medina, M.-C.; Alfaro-Alarcón, A.; Bouza-Mora, L.; Jiménez-Rojas, C.; Wong, M.; Barquero-Calvo, E.; Rojas, N.; et al. *Brucella canis* Is an Intracellular Pathogen That Induces a Lower Proinflammatory Response than Smooth Zoonotic Counterparts. *Infect. Immun.* **2015**, *83*, 4861–4870. [CrossRef]
105. Burberry, A.; Zeng, M.Y.; Ding, L.; Wicks, I.; Inohara, N.; Morrison, S.J.; Núñez, G. Infection Mobilizes Hematopoietic Stem Cells through Cooperative NOD-like Receptor and Toll-like Receptor Signaling. *Cell Host Microbe* **2014**, *15*, 779–791. [CrossRef]
106. Takizawa, H.; Fritsch, K.; Kovtonyuk, L.V.; Saito, Y.; Yakkala, C.; Jacobs, K.; Ahuja, A.K.; Lopes, M.; Hausmann, A.; Hardt, W.-D.; et al. Pathogen-Induced TLR4-TRIF Innate Immune Signaling in Hematopoietic Stem Cells Promotes Proliferation but Reduces Competitive Fitness. *Cell Stem Cell* **2017**, *21*, 225–240.e225. [CrossRef]

107. Hysenaj, L.; De Laval, B.; Arce-Gorvel, V.; Bosilkovski, M.; Gonzalez-Espinoza, G.; Debroas, G.; Sieweke, M.; Sarrazin, S.; Gorvel, J.-P. *Brucella* subverts myeloid commitment of hematopoietic stem cells by a SLAMF1 (CD150)-dependent mechanism. *Cell Stem Cell* **2021**, *26*, 657–674.
108. Demir, C.; Karahocagil, M.K.; Esen, R.; Atmaca, M.; Gönüllü, H.; Akdeniz, H. Bone marrow biopsy findings in brucellosis patients with hematologic abnormalities. *Chin. Med. J.* **2012**, *125*, 1871–1876.
109. Ganado, W.; Bannister, W. Bacteraemia in Human Brucellosis. *BMJ* **1960**, *1*, 601–603. [CrossRef] [PubMed]
110. Gotuzzo, E.; Carrillo, C.; Guerra, J.; Llosa, L. An Evaluation of Diagnostic Methods for Brucellosis—The Value of Bone Marrow Culture. *J. Infect. Dis.* **1986**, *153*, 122–125. [CrossRef] [PubMed]
111. García, P.; Yrivarren, J.L.; Argumans, C.; Crosby, E.; Carrillo, C.; Gotuzzo, E. Evaluation of the bone marrow in patients with brucellosis. Clinico-pathological correlation. *Enferm. Infecc. Y Microbiol. Clin.* **1990**, *8*, 19–24.
112. Crosby, E.; Llosa, L.; Quesada, M.M.; Carrillo, P.C.; Gotuzzo, E. Hematologic Changes in Brucellosis. *J. Infect. Dis.* **1984**, *150*, 419–424. [CrossRef] [PubMed]
113. Ertem, M.; Kürekçi, A.; Aysev, D.; Ünal, E.; Ikincioğulları, A. Brucellosis transmitted by bone marrow transplantation. *Bone Marrow Transplant.* **2000**, *26*, 225–226. [CrossRef] [PubMed]
114. Naparstek, E. Transmission of brucellosis by bone marrow transplantation. *Lancet* **1982**, *319*, 574–575. [CrossRef]
115. Bracewell, C.; Corbel, M. An association between arthritis and persistent serological reactions to *Brucella abortus* in cattle from apparently brucellosis-free herds. *Vet. Rec.* **1980**, *106*, 99–101. [CrossRef] [PubMed]
116. Corbel, M.J.; Stuart, F.A.; Brewer, R.A.; Jeffrey, M.; Bradley, R. Arthropathy associated with *Brucella abortus* strain 19 vaccination in cattle. I. Examination of field cases. *Br. Vet. J.* **1989**, *145*, 337–346. [CrossRef]
117. Delpino, M.V.; Fossati, C.A.; Baldi, P.C. Proinflammatory Response of Human Osteoblastic Cell Lines and Osteoblast-Monocyte Interaction upon Infection with *Brucella* spp. *Infect. Immun.* **2009**, *77*, 984–995. [CrossRef]
118. Khalaf, O.H.; Chaki, S.P.; Garcia-Gonzalez, D.G.; Suva, L.J.; Gaddy, D.; Arenas-Gamboa, A.M. Interaction of *Brucella abortus* with Osteoclasts: A Step toward Understanding Osteoarticular Brucellosis and Vaccine Safety. *Infect. Immun.* **2020**, *88*, e00822-19. [CrossRef]
119. Scian, R.; Barrionuevo, P.; Giambartolomei, G.H.; De Simone, E.A.; Vanzulli, S.I.; Fossati, C.A.; Baldi, P.C.; Delpino, M.V. Potential Role of Fibroblast-Like Synoviocytes in Joint Damage Induced by *Brucella abortus* Infection through Production and Induction of Matrix Metalloproteinases. *Infect. Immun.* **2011**, *79*, 3619–3632. [CrossRef] [PubMed]
120. Giambartolomei, G.H.; Arriola Benitez, P.C.; Delpino, M.V. *Brucella* and Osteoarticular Cell Activation: Partners in Crime. *Front. Microbiol.* **2017**, *8*. [CrossRef]
121. Zhang, J.; Supakorndej, T.; Krambs, J.R.; Rao, M.; Abou-Ezzi, G.; Ye, R.Y.; Li, S.; Trinkaus, K.; Link, D.C. Bone marrow dendritic cells regulate hematopoietic stem/progenitor cell trafficking. *J. Clin. Investig.* **2019**, *129*, 2920–2931. [CrossRef]
122. Winkler, I.G.; Sims, N.A.; Pettit, A.R.; Barbier, V.; Nowlan, B.; Helwani, F.; Poulton, I.J.; van Rooijen, N.; Alexander, K.A.; Raggatt, L.J.; et al. Bone marrow macrophages maintain hematopoietic stem cell (HSC) niches and their depletion mobilizes HSCs. *Blood* **2010**, *116*, 4815–4828. [CrossRef]
123. Casanova-Acebes, M.; Pitaval, C.; Weiss, L.A.; Nombela-Arrieta, C.; Chèvre, R.; A-González, N.; Kunisaki, Y.; Zhang, D.; van Rooijen, N.; Silberstein, L.E.; et al. Rhythmic Modulation of the Hematopoietic Niche through Neutrophil Clearance. *Cell* **2013**, *153*, 1025–1035. [CrossRef]
124. Gomes, A.C.; Saraiva, M.; Gomes, M.S. The bone marrow hematopoietic niche and its adaptation to infection. *Semin. Cell Dev. Biol.* **2020**, S1084952119302617. [CrossRef] [PubMed]
125. Das, B.; Kashino, S.S.; Pulu, I.; Kalita, D.; Swami, V.; Yeger, H.; Felsher, D.W.; Campos-Neto, A. CD271+ Bone Marrow Mesenchymal Stem Cells May Provide a Niche for Dormant *Mycobacterium tuberculosis*. *Sci. Transl. Med.* **2013**, *5*, 170ra113. [CrossRef]
126. Tornack, J.; Reece, S.T.; Bauer, W.M.; Vogelzang, A.; Bandermann, S.; Zedler, U.; Stingl, G.; Kaufmann, S.H.E.; Melchers, F. Human and Mouse Hematopoietic Stem Cells Are a Depot for Dormant *Mycobacterium tuberculosis*. *PLoS ONE* **2017**, *12*, e0169119. [CrossRef] [PubMed]
127. Hardy, J.; Chu, P.; Contag, C.H. Foci of *Listeria monocytogenes* persist in the bone marrow. *Dis. Models Mech.* **2009**, *2*, 39–46. [CrossRef] [PubMed]
128. Wood, E.E. Brucellosis as a Hazard of Blood Transfusion. *BMJ* **1955**, *1*, 27–28. [CrossRef]
129. Gasteiger, G.; Ataide, M.; Kastenmüller, W. Lymph node—An organ for T-cell activation and pathogen defense. *Immunol. Rev.* **2016**, *271*, 200–220. [CrossRef] [PubMed]
130. Krishnamurty, A.T.; Turley, S.J. Lymph node stromal cells: Cartographers of the immune system. *Nat. Immunol.* **2020**, *21*, 369–380. [CrossRef]
131. Jackson, D.G. Leucocyte Trafficking via the Lymphatic Vasculature— Mechanisms and Consequences. *Front. Immunol.* **2019**, *10*, 471. [CrossRef]
132. Lynskey, N.N.; Banerji, S.; Johnson, L.A.; Holder, K.A.; Reglinski, M.; Wing, P.A.C.; Rigby, D.; Jackson, D.G.; Sriskandan, S. Rapid Lymphatic Dissemination of Encapsulated Group A Streptococci via Lymphatic Vessel Endothelial Receptor-1 Interaction. *PLoS Pathog.* **2015**, *11*, e1005137. [CrossRef]
133. MacArthur, W.P. The Nature of Brucellosis. *Am. J. Trop. Med. Hyg.* **1957**, *6*, 591–592. [CrossRef]
134. Fensterbank, R. Some aspects of experimental bovine brucellosis. *Ann. Rech. Vet. Vet. Ann. Vet. Res.* **1987**, *18*, 421–428.

135. Meador, V.P.; Deyoe, B.L.; Cheville, N.F. Effect of Nursing on *Brucella abortus* Infection of Mammary Glands of Goats. *Vet. Pathol.* **1989**, *26*, 369–375. [CrossRef] [PubMed]
136. Meador, V.P.; Deyoe, B.L.; Cheville, N.F. Pathogenesis of *Brucella abortus* Infection of the Mammary Gland and Supramammary Lymph Node of the Goat. *Vet. Pathol.* **1989**, *26*, 357–368. [CrossRef] [PubMed]
137. Scientific Committe on Animal Health and Animal Welfare. Brucellosis in Sheep and Goats (Brucella melitensis). 2001. Available online: https://ec.europa.eu/food/sites/food/files/safety/docs/sci-com_scah_out59_en.pdf (accessed on 15 December 2020).
138. Damir, H.A.; Tageldin, M.H.; Kenyon, S.J.; Idris, O.F. Isolation of *Brucella abortus* from experimentally infected dromedary camels in Sudan: A preliminary report. *Vet. Res. Commun.* **1989**, *13*, 403–406. [CrossRef]
139. Foster, J.T.; Walker, F.M.; Rannals, B.D.; Hussain, M.H.; Drees, K.P.; Tiller, R.V.; Hoffmaster, A.R.; Al-Rawahi, A.; Keim, P.; Saqib, M. African Lineage *Brucella melitensis* Isolates from Omani Livestock. *Front. Microbiol.* **2018**, *8*, 2702. [CrossRef]
140. Marin, C.M.; Jimenez de Bagues, M.P.; Barberan, M.; Blasco, J.M. Comparison of two selective media for the isolation of *Brucella melitensis* from naturally infected sheep and goats. *Vet. Rec.* **1996**, *138*, 409–411. [CrossRef]
141. Cruz-Migoni, S.; Caamaño, J. Fat-Associated Lymphoid Clusters in Inflammation and Immunity. *Front. Immunol.* **2016**, *7*. [CrossRef]
142. Ferrante, A.W. The immune cells in adipose tissue. *Diabetes Obes. Metab.* **2013**, *15*, 34–38. [CrossRef]
143. Bechah, Y.; Paddock, C.D.; Capo, C.; Mege, J.-L.; Raoult, D. Adipose Tissue Serves as a Reservoir for Recrudescent *Rickettsia prowazekii* Infection in a Mouse Model. *PLoS ONE* **2010**, *5*, e8547. [CrossRef]
144. Beigier-Bompadre, M.; Montagna, G.N.; Kühl, A.A.; Lozza, L.; Weiner, J.; Kupz, A.; Vogelzang, A.; Mollenkopf, H.-J.; Löwe, D.; Bandermann, S.; et al. *Mycobacterium tuberculosis* infection modulates adipose tissue biology. *PLoS Pathog.* **2017**, *13*, e1006676. [CrossRef]
145. Cabalén, M.E.; Cabral, M.F.; Sanmarco, L.M.; Andrada, M.C.; Onofrio, L.I.; Ponce, N.E.; Aoki, M.P.; Gea, S.; Cano, R.C. Chronic *Trypanosoma cruzi* infection potentiates adipose tissue macrophage polarization toward an anti-inflammatory M2 phenotype and contributes to diabetes progression in a diet-induced obesity model. *Oncotarget* **2016**, *7*, 13400–13415. [CrossRef]
146. Damouche, A.; Lazure, T.; Avettand-Fènoël, V.; Huot, N.; Dejucq-Rainsford, N.; Satie, A.-P.; Mélard, A.; David, L.; Gommet, C.; Ghosn, J.; et al. Adipose Tissue Is a Neglected Viral Reservoir and an Inflammatory Site during Chronic HIV and SIV Infection. *PLoS Pathog.* **2015**, *11*, e1005153. [CrossRef]
147. Neyrolles, O.; Hernández-Pando, R.; Pietri-Rouxel, F.; Fornès, P.; Tailleux, L.; Payán, J.A.B.; Pivert, E.; Bordat, Y.; Aguilar, D.; Prévost, M.-C.; et al. Is Adipose Tissue a Place for *Mycobacterium tuberculosis* Persistence? *PLoS ONE* **2006**, *1*, e43. [CrossRef] [PubMed]
148. Figueiredo, L. The Parasite in the Fat: A Serendipitous Finding. *Trends Parasitol.* **2017**, *33*, 255. [CrossRef]
149. De Souza, T.D.; de Carvalho, T.F.; Mol, J.P.d.S.; Lopes, J.V.M.; Silva, M.F.; da Paixão, T.A.; Santos, R.L. Tissue distribution and cell tropism of *Brucella canis* in naturally infected canine foetuses and neonates. *Sci. Rep.* **2018**, *8*, 7203. [CrossRef] [PubMed]
150. Pesce Viglietti, A.I.; Giambartolomei, G.H.; Quarleri, J.; Delpino, M.V. *Brucella abortus* Infection Modulates 3T3-L1 Adipocyte Inflammatory Response and Inhibits Adipogenesis. *Front. Endocrinol.* **2020**, *11*, 585923. [CrossRef] [PubMed]
151. Shi, Y.; Gao, H.; Pappas, G.; Chen, Q.; Li, M.; Xu, J.; Lai, S.; Liao, Q.; Yang, W.; Yi, Z.; et al. Clinical features of 2041 human brucellosis cases in China. *PLoS ONE* **2018**, *13*, e0205500. [CrossRef]
152. Hansel, W. The Essentiality of the Epididymal Fat Pad for Spermatogenesis. *Endocrinology* **2010**, *151*, 5565–5567. [CrossRef] [PubMed]
153. Elewa, Y.H.A.; Ichii, O.; Otsuka, S.; Hashimoto, Y.; Kon, Y. Characterization of mouse mediastinal fat-associated lymphoid clusters. *Cell Tissue Res.* **2014**, *357*, 731–741. [CrossRef]
154. Buscher, K.; Wang, H.; Zhang, X.; Striewski, P.; Wirth, B.; Saggu, G.; Lütke-Enking, S.; Mayadas, T.N.; Ley, K.; Sorokin, L.; et al. Protection from septic peritonitis by rapid neutrophil recruitment through omental high endothelial venules. *Nat. Commun.* **2016**, *7*, 10828. [CrossRef]
155. Cipolletta, D.; Cohen, P.; Spiegelman, B.M.; Benoist, C.; Mathis, D. Appearance and disappearance of the mRNA signature characteristic of T_{reg} cells in visceral adipose tissue: Age, diet, and PPARγ effects. *Proc. Natl. Acad. Sci. USA* **2015**, *112*, 482–487. [CrossRef] [PubMed]
156. Meza-Perez, S.; Randall, T.D. Immunological Functions of the Omentum. *Trends Immunol.* **2017**, *38*, 526–536. [CrossRef]
157. Xavier, M.N.; Winter, M.G.; Spees, A.M.; Nguyen, K.; Atluri, V.L.; Silva, T.M.A.; Bäumler, A.J.; Müller, W.; Santos, R.L.; Tsolis, R.M. CD4+ T Cell-derived IL-10 Promotes *Brucella abortus* Persistence via Modulation of Macrophage Function. *PLoS Pathog.* **2013**, *9*, e1003454. [CrossRef] [PubMed]
158. Garin-Bastuji, B.; Hars, J.; Drapeau, A.; Cherfa, M.-A.; Game, Y.; Le Horgne, J.-M.; Rautureau, S.; Maucci, E.; Pasquier, J.-J.; Jay, M.; et al. Reemergence of *Brucella melitensis* Infection in Wildlife, France. *Emerg. Infect. Dis.* **2014**, *20*, 1570–1571. [CrossRef] [PubMed]
159. Lambert, S.; Gilot-Fromont, E.; Freycon, P.; Thébault, A.; Game, Y.; Toïgo, C.; Petit, E.; Barthe, M.-N.; Reynaud, G.; Jaÿ, M.; et al. High Shedding Potential and Significant Individual Heterogeneity in Naturally-Infected Alpine ibex (*Capra ibex*) With *Brucella melitensis*. *Front. Microbiol.* **2018**, *9*, 1065. [CrossRef] [PubMed]
160. Banskar, S.; Bhute, S.S.; Suryavanshi, M.V.; Punekar, S.; Shouche, Y.S. Microbiome analysis reveals the abundance of bacterial pathogens in *Rousettus leschenaultii* guano. *Sci. Rep.* **2016**, *6*, 36948. [CrossRef]
161. Bai, Y.; Urushadze, L.; Osikowicz, L.; McKee, C.; Kuzmin, I.; Kandaurov, A.; Babuadze, G.; Natradze, I.; Imnadze, P.; Kosoy, M. Molecular Survey of Bacterial Zoonotic Agents in Bats from the Country of Georgia (Caucasus). *PLoS ONE* **2017**, *12*, e0171175. [CrossRef]

162. Shirima, G.M.; Kunda, J.S. Prevalence of brucellosis in the human, livestock and wildlife interface areas of Serengeti National Park, Tanzania. *Onderstepoort J. Vet. Res.* **2016**, *83*, a1032. [CrossRef]
163. Tessaro, S.V.; Forbes, L.B.; Turcotte, C. A survey of brucellosis and tuberculosis in bison in and around Wood Buffalo National Park, Canada. *Can. Vet. J. La Rev. Vet. Can.* **1990**, *31*, 174–180.
164. Kreizinger, Z.; Foster, J.T.; Rónai, Z.; Sulyok, K.M.; Wehmann, E.; Jánosi, S.; Gyuranecz, M. Genetic relatedness of *Brucella suis* biovar 2 isolates from hares, wild boars and domestic pigs. *Vet. Microbiol.* **2014**, *172*, 492–498. [CrossRef]
165. Forbes, L.B.; Tessaro, S.V. Transmission of brucellosis from reindeer to cattle. *J. Am. Vet. Med. Assoc.* **1993**, *203*, 289–294.
166. Foster, G.; Jahans, K.L.; Reid, R.J.; Ross, H.M. Isolation of *Brucella* species from cetaceans, seals and an otter. *Vet. Rec.* **1996**, *138*, 583–586. [CrossRef]
167. Scholz, H.C.; Hubalek, Z.; Sedlacek, I.; Vergnaud, G.; Tomaso, H.; Al Dahouk, S.; Melzer, F.; Kampfer, P.; Neubauer, H.; Cloeckaert, A.; et al. *Brucella microti* sp. nov., isolated from the common vole *Microtus arvalis*. *Int. J. Syst. Evol. Microbiol.* **2008**, *58*, 375–382. [CrossRef] [PubMed]
168. Scholz, H.C.; Hofer, E.; Vergnaud, G.; Fleche, P.L.; Whatmore, A.M.; Dahouk, S.A.; Pfeffer, M.; Krüger, M.; Cloeckaert, A.; Tomaso, H. Isolation of *Brucella microti* from Mandibular Lymph Nodes of Red Foxes, *Vulpes vulpes*, in Lower Austria. *Vector Borne Zoonotic Dis.* **2009**, *9*, 153–156. [CrossRef]
169. Kimura, M.; Une, Y.; Suzuki, M.; Park, E.-S.; Imaoka, K.; Morikawa, S. Isolation of *Brucella inopinata* -Like Bacteria from White's and Denny's Tree Frogs. *Vector Borne Zoonotic Dis.* **2017**, *17*, 297–302. [CrossRef] [PubMed]
170. Dawson, C.E.; Perrett, L.L.; Stubberfield, E.J.; Stack, J.A.; Farrelly, S.S.; Cooley, W.A.; Davison, N.J.; Quinney, S. Isolation and characterization of *Brucella* from the lungworms of a harbor porpoise (*Phocoena phocoena*). *J. Wildl. Dis.* **2008**, *44*, 237–246. [CrossRef] [PubMed]
171. Eisenberg, T.; Riße, K.; Schauerte, N.; Geiger, C.; Blom, J.; Scholz, H.C. Isolation of a novel 'atypical' *Brucella* strain from a bluespotted ribbontail ray (*Taeniura lymma*). *Antonie Van Leeuwenhoek* **2017**, *110*, 221–234. [CrossRef]
172. Eisenberg, T.; Schlez, K.; Fawzy, A.; Völker, I.; Hechinger, S.; Curić, M.; Schauerte, N.; Geiger, C.; Blom, J.; Scholz, H.C. Expanding the host range: Infection of a reptilian host (*Furcifer pardalis*) by an atypical *Brucella* strain. *Antonie Van Leeuwenhoek* **2020**, *113*, 1531–1537. [CrossRef] [PubMed]
173. Jiménez de Bagüés, M.P.; Ouahrani-Bettache, S.; Quintana Juan, F.; Mitjana, O.; Hanna, N.; Bessoles, S.; Sanchez, F.; Scholz, H.C.; Lafont, V.; Köhler, S.; et al. The New Species *Brucella microti* Replicates in Macrophages and Causes Death in Murine Models of Infection. *J. Infect. Dis.* **2010**, *202*, 3–10. [CrossRef]
174. Hernández-Mora, G.; Palacios-Alfaro, J.D.; González-Barrientos, R. Wildlife reservoirs of brucellosis: *Brucella* in aquatic environments. *Rev. Sci. Tech.* **2013**, *32*, 89–103. [CrossRef] [PubMed]
175. Garner, M.M.; Lambourn, D.M.; Jeffries, S.J.; Hall, P.B.; Rhyan, J.C.; Ewalt, D.R.; Polzin, L.M.; Cheville, N.F. Evidence of *Brucella* infection in Parafilaroides lungworms in a Pacific harbor seal (*Phoca vitulina richardsi*). *J. Vet. Diagn. Investig.* **1997**, *9*, 298–303. [CrossRef] [PubMed]
176. Perrett, L.L.; Dawson, C.E.; Davison, N.; Quinney, S. *Brucella* infection of lungworms from a harbour porpoise. *Vet. Rec.* **2004**, *154*, 800. [PubMed]
177. Lambourn, D.M.; Garner, M.; Ewalt, D.; Raverty, S.; Sidor, I.; Jeffries, S.J.; Rhyan, J.; Gaydos, J.K. *Brucella pinnipedialis* infections in Pacific harbor seals (*Phoca vitulina richardsi*) from Washington State, USA. *J. Wildl. Dis.* **2013**, *49*, 802–815. [CrossRef] [PubMed]
178. Nymo, I.H.; Seppola, M.; Al Dahouk, S.; Bakkemo, K.R.; Jiménez de Bagüés, M.P.; Godfroid, J.; Larsen, A.K. Experimental Challenge of Atlantic Cod (*Gadus morhua*) with a *Brucella pinnipedialis* Strain from Hooded Seal (*Cystophora cristata*). *PLoS ONE* **2016**, *11*, e0159272. [CrossRef] [PubMed]
179. El-Tras, W.F.; Tayel, A.A.; Eltholth, M.M.; Guitian, J. *Brucella* infection in fresh water fish: Evidence for natural infection of Nile catfish, *Clarias gariepinus*, with *Brucella melitensis*. *Vet. Microbiol.* **2010**, *141*, 321–325. [CrossRef] [PubMed]
180. Tiller, R.V.; Gee, J.E.; Lonsway, D.R.; Gribble, S.; Bell, S.C.; Jennison, A.V.; Bates, J.; Coulter, C.; Hoffmaster, A.R.; De, B.K. Identification of an unusual *Brucella* strain (BO2) from a lung biopsy in a 52 year-old patient with chronic destructive pneumonia. *BMC Microbiol.* **2010**, *10*, 23. [CrossRef]
181. Soler-Lloréns, P.F.; Quance, C.R.; Lawhon, S.D.; Stuber, T.P.; Edwards, J.F.; Ficht, T.A.; Robbe-Austerman, S.; O'Callaghan, D.; Keriel, A. A *Brucella* spp. Isolate from a Pac-Man Frog (*Ceratophrys ornata*) Reveals Characteristics Departing from Classical Brucellae. *Front. Cell. Infect. Microbiol.* **2016**, *6*. [CrossRef]
182. Rouzic, N.; Desmier, L.; Cariou, M.-E.; Gay, E.; Foster, J.T.; Williamson, C.H.D.; Schmitt, F.; Le Henaff, M.; Le Coz, A.; Lorléac'h, A.; et al. First Case of Brucellosis Caused by an Amphibian-type *Brucella*. *Clin. Infect. Dis.* **2020**. [CrossRef] [PubMed]
183. Ponsart, C.; Riou, M.; Locatelli, Y.; Jacques, I.; Fadeau, A.; Jay, M.; Simon, R.; Perrot, L.; Freddi, L.; Breton, S.; et al. *Brucella melitensis* Rev.1 vaccination generates a higher shedding risk of the vaccine strain in Alpine ibex (*Capra ibex*) compared to the domestic goat (*Capra hircus*). *Vet. Res.* **2019**, *50*, 100. [CrossRef]
184. Eric, J.M.; Brandon, M.S.; Paul, C.C.; Jared, D.R.; William, H.E.; Benjamin, W.; Scott, G.S.; Terry, J.K. Assessment of a Strain 19 Brucellosis Vaccination Program in Elk. *Wildl. Soc. Bull.* **2017**, *41*, 70–79. [CrossRef]
185. Davis, D.S.; Elzer, P.H. *Brucella* vaccines in wildlife. *Vet. Microbiol.* **2002**, *90*, 533–544. [CrossRef]
186. Godfroid, J.; Nielsen, K.; Saegerman, C. Diagnosis of brucellosis in livestock and wildlife. *Croat. Med. J.* **2010**, *51*, 296–305. [CrossRef] [PubMed]

187. Dadar, M.; Shahali, Y.; Fakhri, Y.; Godfroid, J. The global epidemiology of *Brucella* infections in terrestrial wildlife: A meta-analysis. *Transbound. Emerg. Dis.* **2020**, tbed.13735. [CrossRef] [PubMed]
188. Batista, T.G.D.S.; Fornazari, F.; Joaquim, S.F.; Latosinski, G.S.; Teixeira, C.R.; Langoni, H. Serologic Screening for Smooth *Brucella* sp. in Wild Animals in Brazil. *J. Wildl. Dis.* **2019**, *55*, 721–723. [CrossRef] [PubMed]

MDPI
St. Alban-Anlage 66
4052 Basel
Switzerland
Tel. +41 61 683 77 34
Fax +41 61 302 89 18
www.mdpi.com

Pathogens Editorial Office
E-mail: pathogens@mdpi.com
www.mdpi.com/journal/pathogens

www.ingramcontent.com/pod-product-compliance
Lightning Source LLC
Chambersburg PA
CBHW041951260926
51850CB00019B/2532
9783036524184